High-Yield Embryology

Look for these published and forthcoming High-Yield™ titles:

Published

Fix, **High-Yield™ Neuroanatomy**
Glaser, **High-Yield™ Biostatistics**

Forthcoming

Fadem, **High-Yield™ Behavioral Science**
Dudek, **High-Yield™ Histology**
Dudek, **High-Yield™ Gross Anatomy**

High-Yield Embryology

Ronald W. Dudek, Ph.D.
Department of Anatomy and Cell Biology
East Carolina University
School of Medicine
Greenville, North Carolina

LIPPINCOTT WILLIAMS & WILKINS
A **Wolters Kluwer** Company
Philadelphia • Baltimore • New York • London
Buenos Aires • Hong Kong • Sydney • Tokyo

Editor: Elizabeth Nieginski
Managing Editor: Alethea H. Elkins
Development Editor: Melanie Cann
Production Coordinator: Felecia R. Weber
Illustrator: Michael Malicki
Cover Designer: Karen Klinedinst
Typesetter: Maryland Composition
Printer: Victor Graphics
Binder: Victor Graphics

351 West Camden Street
Baltimore, Maryland 21201-2436 USA

Rose Tree Corporate Center
1400 North Providence Road
Building II, Suite 5025
Media, Pennsylvania 19063-2043 USA

Printed in the United States of America

First Edition,

Library of Congress Cataloging in Publication Data

Dudek, Ronald W., 1950–
High-yield embryology / Ronald W. Dudek.
p. cm.
ISBN 0-683-02714-X. — ISBN 0-683-02714-X
1. Embryology, Human—Outlines, syllabi, etc. I. Title.
[DNLM: 1. Embryology—examination questions. 2. Embryo—anatomy & histology—examination questions. 3. Embryo—physiology—examination questions. 4. Fetal Development—examination questions. QS 618.2 D845h 1996]
QM601.D83 1996
612.6′4′076—dc20
DNLM/DLC
for Library of Congress 96-5667
CIP

To purchase additional copies of this book, call our customer service department at **(800) 638-0672** or fax orders to **(800) 447-8438.** For other book services, including chapter reprints and large quantity sales, ask for the Special Sales Department.

Canadian customers should call **(800) 268-4178,** or fax **(905) 470-6780.** For all other calls originating outside of the United States, please call **(410) 528-4223** or fax us at **(410) 528-8550.**

Visit Williams & Wilkins on the Internet: **http://www.wwilkins.com** or contact our customer service department at **custserv@wwilkins.com.** Williams & Wilkins customer service representatives are available from 8:30 am to 6:00 pm, EST, Monday through Friday, for telephone access.

99
4 5 6 7 8 9 10

Contents

Preface

High-Yield Embryology reviews those themes in the discipline of embryology that have a high likelihood of appearing on the national board exams. This book assumes that the student has a solid background in Embryology.

National board exam questions in embryology are frequently concerned with the embryologic derivation of adult structures or remnants. Tables are included where appropriate to aid the student in easy review. Illustrations give the student some visual reinforcement of the area under consideration. In addition, clinical correlations are provided for the most common congenital malformations.

Dedication

I would like to dedicate this book to my father, Stanley J. Dudek, who died Sunday, March 20, 1988, at 11 A.M. It was his hard work and sacrifice that allowed me access to the finest educational institutions in the country. It was by hard work and sacrifice that he showed his love for his wife, Lottie; daughter, Christine; and grandchildren, Karolyn, Jeannie, and Katie. I remember my father as a good man who did the best he could. He is missed and remembered.

1

Prefertilization Events

I. OVERVIEW. Gametes **(oocytes** and **spermatozoa),** descendants of **primordial germ cells,** are produced in the adult by either **oogenesis** or **spermatogenesis,** processes that involve **meiosis.** Primordial germ cells originate in the **wall of the yolk sac** of the embryo and migrate to the gonadal region.

II. MEIOSIS (Figure 1-1), which occurs **only during the production of gametes,** consists of two cell divisions (**meiosis I** and **meiosis II**) and results in the formation of four gametes containing 23 chromosomes and 1N amount of DNA (23,1N). Meiosis:

A. Reduces the number of chromosomes within gametes to ensure that the human species number of chromosomes (46) is maintained from generation to generation

B. Redistributes maternal and paternal chromosomes to insure genetic variability

C. Promotes the exchange of small amounts of maternal and paternal DNA via **crossover**

III. FEMALE GAMETOGENESIS (OOGENESIS; Table 1-1)

A. Primordial germ cells (46,2N) arrive in the ovary at week 4 of embryonic development and differentiate into **oogonia (46,2N).**

B. Oogonia enter **meiosis I** and undergo DNA replication to form **primary oocytes (46,4N).** All primary oocytes are **formed by the fifth month of fetal life** and **remain dormant in prophase (dictyotene) of meiosis I until puberty.**

C. During a woman's ovarian cycle, a primary oocyte completes meiosis I to form a **secondary oocyte (23,2N)** and a **first polar body,** which probably degenerates.

D. The secondary oocyte enters **meiosis II** and ovulation occurs when the chromosomes align at metaphase. The secondary oocyte remains **arrested in metaphase of meiosis II** until fertilization occurs.

E. At fertilization, the secondary oocyte completes meiosis II to form a **mature oocyte (23,1N)** and a **second polar body.**

IV. MALE GAMETOGENESIS

A. Spermatogenesis

1. **Primordial germ cells (46,2N)** arrive in the testes at week 4 of embryonic development and remain dormant until puberty. At puberty, primordial germ cells differentiate into **type A spermatogonia (46,2N).**

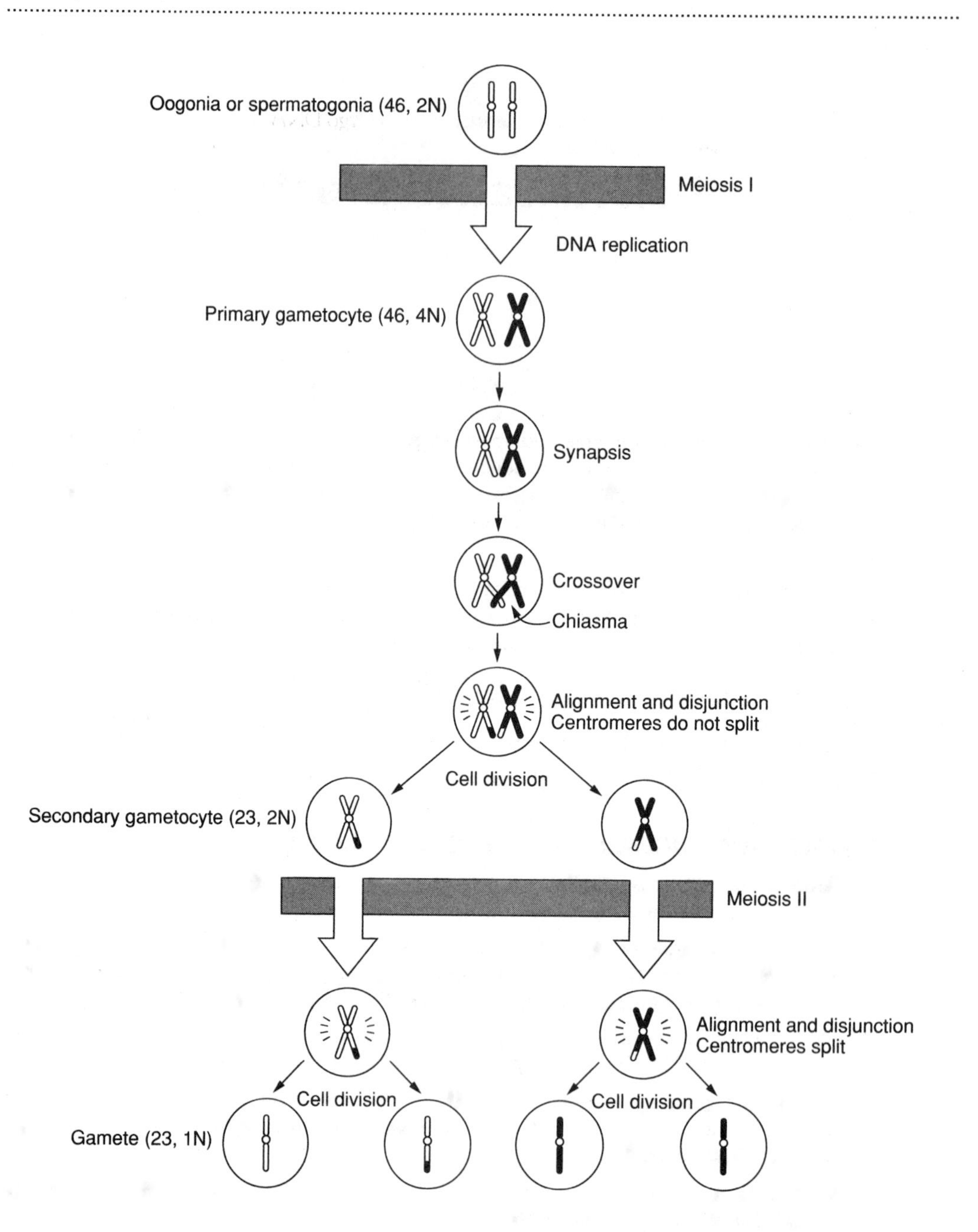

Figure 1-1. Meiosis. Note that only one pair of homologous chromosomes is shown (*white* = maternal origin; *black* = paternal origin). Synapsis is the process of pairing of homologous chromosomes. The point at which the DNA molecule crosses over, the chiasma, is where the exchange of small amounts of maternal and paternal DNA takes place. (Modified with permission from Fix JD and Dudek RW: *BRS Embryology*, Baltimore, Williams & Wilkins, 1995, p 4.)

2. Type A spermatogonia undergo **mitosis** to form either more type A spermatogonia (to maintain the supply) or **type B spermatogonia (46,2N).**
3. Type B spermatogonia enter meiosis I and undergo DNA replication to form **primary spermatocytes (46,4N).**
4. Primary spermatocytes complete meiosis I to form **two secondary spermatocytes (23,2N).**
5. Secondary spermatocytes complete meiosis II to form **four spermatids (23,1N).**

B. Spermiogenesis is a post-meiotic series of morphological changes within the spermatid that results in the formation of the head, body, and tail of the **sperm (23,1N).**

C. Capacitation. Newly ejaculated sperm are incapable of fertilization until they undergo capacitation, which occurs in the female reproductive tract and involves the unmasking of sperm glycosyltransferases and removal of the proteins that coat the surface of the sperm.

Table 1-1

Number of Chromosomes and Amount of DNA Contained in Cells During the Stages of Gametogenesis

Cell	Number of Chromosomes Amount of DNA
Primordial germ cell, oogonia,spermatogonia (types A and B)	46,2N
Primary oocyte, primary spermatocyte	46,4N
Secondary oocyte, secondary spermatocyte	23,2N
Oocyte (ovam), spermatid, sperm	23,1N

2

Week 1 (Days 1–7)*

I. OVERVIEW. Figure 2-1 summarizes the events that occur during week 1, following fertilization.

II. FERTILIZATION occurs in the **ampulla of the uterine tube.**

A. The sperm binds to the zona pellucida of the secondary oocyte and triggers the **acrosome reaction,** causing the release of acrosomal enzymes (e.g., **acrosin**).

B. Aided by the acrosomal enzymes, the sperm penetrates the zona pellucida. Penetration of the zona pellucida elicits the **cortical reaction,** rendering the secondary oocyte **impermeable to other sperm.**

C. The sperm and secondary oocyte cell membranes fuse, and the contents of the sperm enter the cytoplasm of the oocyte.

1. The male genetic material forms the **male pronucleus.**

2. The tail and mitochondria of the sperm degenerate. Therefore, all mitochondria within the zygote are of maternal origin (i.e., **all mitochondrial DNA is of maternal origin**).

D. The secondary oocyte completes meiosis II, forming a mature **ovum.** The nucleus of the ovum is the **female pronucleus.**

E. The male and female pronuclei fuse to form a **zygote.**

III. CLEAVAGE is a series of **mitotic** divisions of the zygote.

A. The zygote cytoplasm is successively cleaved to form a **blastula** consisting of increasingly smaller **blastomeres** (e.g., the first blastomere stage consists of 2 cells; the next, 4 cells; the next, 8 cells).

B. At the 32-cell stage, the blastomeres form a **morula** consisting of an **inner cell mass** and an **outer cell mass.**

* The age of the developing conceptus can be measured either from the estimated day of fertilization (**fertilization age**) or from the day of the **last normal menstrual period (LNMP).** In this book, ages are presented as the fertilization age.

IV. BLASTOCYST FORMATION occurs when fluid secreted within the morula forms the **blastocyst cavity.**

A. The **inner cell mass,** which will eventually become the embryo and fetus, is designated the **embryoblast.**

B. The **outer cell mass,** which becomes part of the **placenta,** is designated the **trophoblast.**

V. IMPLANTATION

A. The **zona pellucida must degenerate** for implantation to occur.

B. The blastocyst implants within the **functional layer** of the endometrium during the **secretory phase** of the menstrual cycle.

1. Implantation usually occurs on the **posterior superior wall** of the uterus.
2. **Ectopic tubal pregnancy** occurs when the blastocyst implants within the uterine tube.

C. The trophoblast differentiates into the **cytotrophoblast** and **syncytiotrophoblast.**

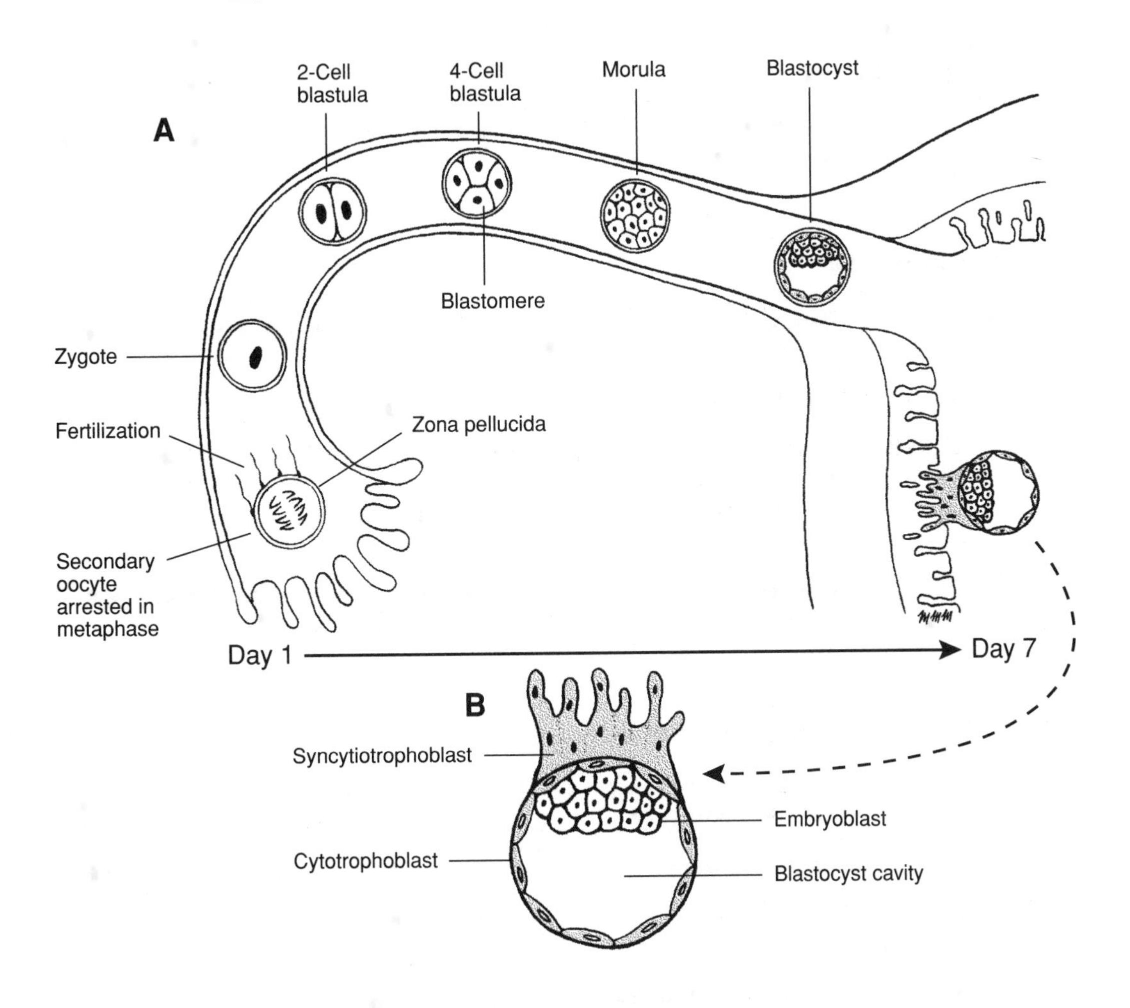

Figure 2-1. (A) The stages of human development during week 1. (B) A day 7 blastocyst.

3

Week 2 (Days 8–14)

I. EMBRYOBLAST

I. EMBRYOBLAST (Figure 3-1). The embryoblast differentiates into two distinct cell layers, the **epiblast** and **hypoblast,** forming a **bilaminar embryonic disk.**

A. Epiblast. Clefts develop within the epiblast to form the **amniotic cavity.**

B. Hypoblast cells migrate along the cytotrophoblast, forming the **yolk sac.**

C. The **prochordal plate,** formed from the fusion of hypoblast and epiblast cells, marks the site of the future **mouth.**

II. TROPHOBLAST

A. The **syncytiotrophoblast** continues its growth into the endometrium to make contact with endometrial blood vessels and glands.

1. The syncytiotrophoblast **does not divide mitotically.**
2. The syncytiotrophoblast produces **human chorionic gonadotropin (hCG),** a glycoprotein that stimulates the production of progesterone in the corpus luteum. hCG can be assayed in **maternal blood** or **urine** at day 10 and is the basis of early pregnancy diagnosis.

B. The **cytotrophoblast** does divide mitotically, adding to the growth of the syncytiotrophoblast. **Primary chorionic villi** protrude into the syncytiotrophoblast.

III. EXTRAEMBRYONIC MESODERM

III. EXTRAEMBRYONIC MESODERM is a new layer of cells derived from the **epiblast.**

A. Extraembryonic somatic mesoderm lines the cytotrophoblast, forms the connecting stalk, and covers the amnion (see Figure 3-1).

1. The conceptus is suspended by the connecting stalk within the **chorionic cavity.**
2. The wall of the chorionic cavity is called the **chorion** and consists of the extraembryonic somatic mesoderm, the cytotrophoblast, and the syncytiotrophoblast.

B. Extraembryonic visceral mesoderm covers the **yolk sac.**

IV. CLINICAL CORRELATIONS

A. Hydatidiform mole. A blighted blastocyst leads to death of the embryo, which is followed by hyperplastic proliferation of the trophoblast within the uterine wall.

B. Choriocarcinoma is a malignant tumor of the trophoblast that may occur following a normal pregnancy, abortion, or a hydatidiform mole.

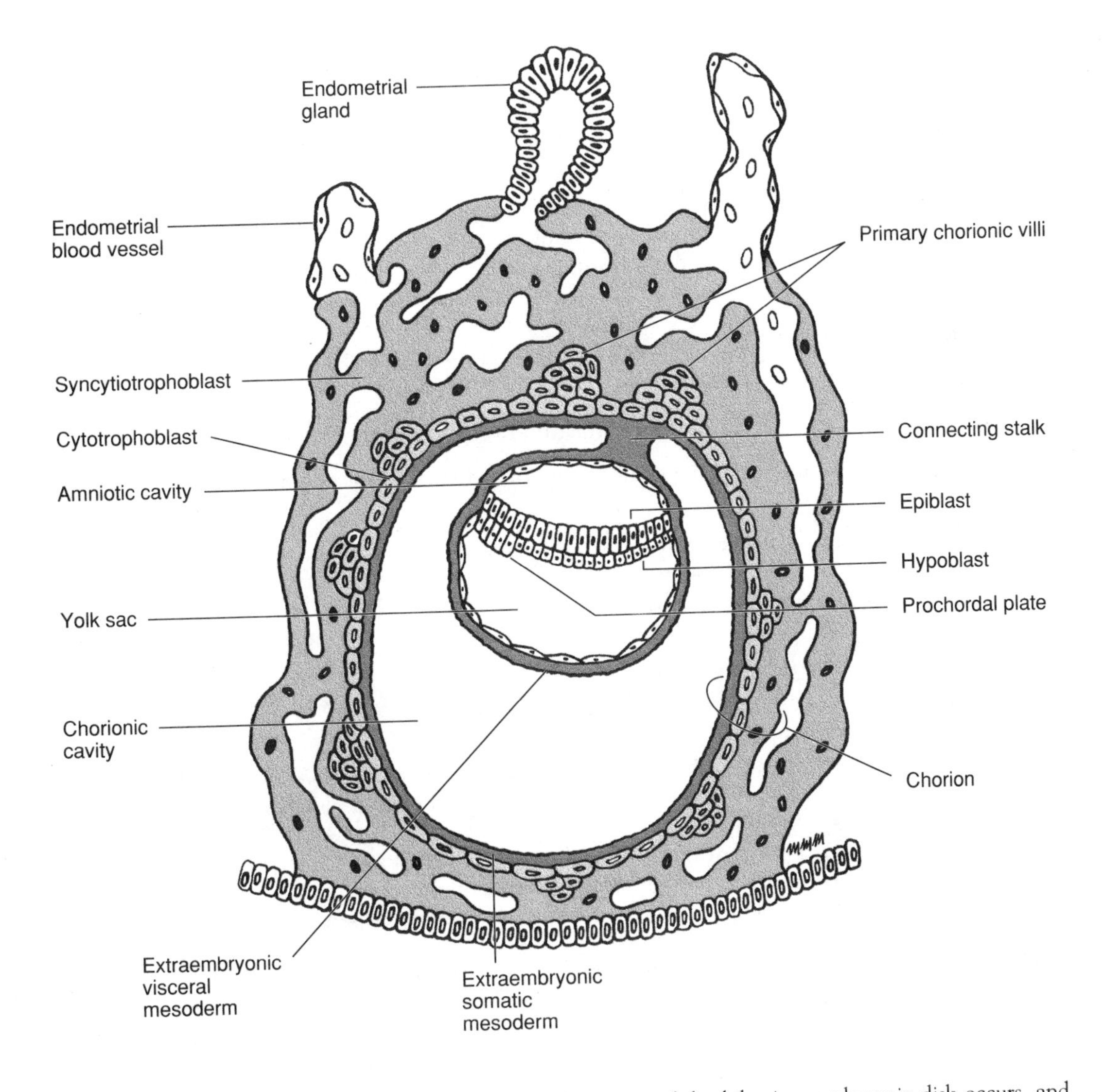

Figure 3-1. A day 14 blastocyst. At this stage, formation of the bilaminar embryonic disk occurs, and implantation within the endometrium is completed.

4

Embryonic Period (Weeks 3–8)

I. INTRODUCTION. All major organ systems begin to develop during the embryonic period, causing a craniocaudal and lateral body folding of the embryo. By the end of the embryonic period (week 8), the embryo has a distinct human appearance.

II. GASTRULATION (Figure 4-1) is the process that establishes the three primary germ layers: **ectoderm, mesoderm,** and **endoderm.**

A. Gastrulation is initiated by the formation of the **primitive streak** within the epiblast.

B. All cells and tissues of the adult can be traced back to the three primary germ layers.

1. Ectoderm

a. Surface ectoderm

(1) Lens of the eye
(2) Adenohypophysis
(3) Utricle, semicircular ducts, and vestibular ganglion of cranial nerve (CN) VIII
(4) Saccule, cochlear duct (organ of Corti), and spiral ganglion of CN VIII
(5) Epithelial lining of the external auditory meatus
(6) Olfactory placode, including CN I
(7) Epithelial lining of the anterior two thirds of the tongue, the hard palate, sides of the mouth, ameloblasts, and parotid glands and ducts
(8) Mammary glands
(9) Epithelial lining of the lower anal canal
(10) Epithelial lining of the distal penile urethra
(11) Epidermis, hair, nails, sweat and cutaneous sebaceous glands

b. Neuroectoderm

(1) All neurons within the central nervous system (CNS), including the preganglionic sympathetic and preganglionic parasympathetic neurons
(2) Astrocytes, oligodendrocytes, ependymocytes, tanycytes, choroid plexus cells
(3) Retina
(4) Pineal gland
(5) Neurohypophysis

c. Neural crest

(1) Postganglionic sympathetic neurons within the sympathetic chain ganglia and prevertebral ganglia

(2) Postganglionic parasympathetic neurons within the ciliary, pterygopalatine, submandibular, otic, enteric ganglia, and ganglia of the abdominal and pelvic cavities
(3) Sensory neurons within the dorsal root ganglia
(4) Schwann cells
(5) Pia mater and arachnoid membrane
(6) Chromaffin cells of the adrenal medulla
(7) Melanocytes
(8) Maxilla, zygomatic bone, temporal bone, palatine bone, vomer, mandible, hard palate, incus, malleus, stapes, sphenomandibular ligament, styloid process, stylohyoid ligament, hyoid bone, frontal bone, parietal bone, sphenoid bone, and ethmoid bone

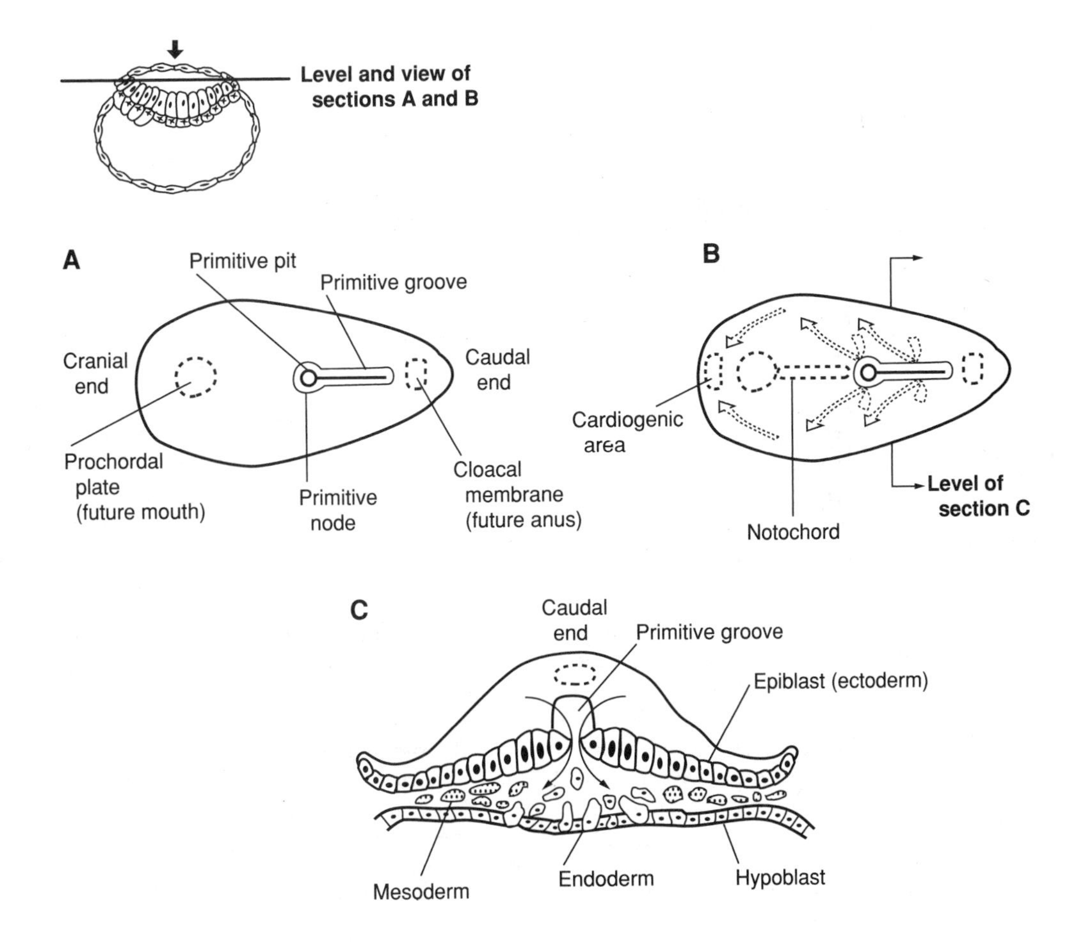

Figure 4-1. Gastrulation. The embryoblast in the upper left-hand corner is provided for orientation. (A) Dorsal view of the epiblast. The primitive streak consists of the primitive groove, node, and pit. (B) *Arrows* show the migration of cells through the primitive streak. The notochord (i.e., mesoderm located between the primitive node and the prochordal plate) induces the formation of the neural tube. The cardiogenic area is the future heart. (C) Epiblast cells migrate to the primitive streak and insert themselves between the epiblast and the hypoblast. Some epiblast cells displace the hypoblast to form endoderm; the remainder migrate cranially, laterally, and along the midline to form mesoderm. After gastrulation, the epiblast is called ectoderm. (Modified with permission from Fix JD and Dudek RW: *BRS Embryology*, Baltimore, Williams & Wilkins, 1995, p 30.)

(9) Odontoblasts
(10) Aorticopulmonary septum
(11) Parafollicular cells
(12) Dilator and sphincter pupillae muscles
(13) Ciliary muscle
(14) Carotid body

2. **Mesoderm**
   a. **Paraxial** (35 pairs of somites and somitomeres)
      (1) Skeletal muscles of trunk
      (2) Skeletal muscles of head and neck
      (3) Extraocular muscles
      (4) Intrinsic muscles of the tongue
      (5) Vertebrae and ribs
      (6) Occipital bone
      (7) Dermis
      (8) Dura mater
   b. **Intermediate**
      (1) Kidneys
      (2) Testes and ovaries
      (3) Genital ducts and accessory sex glands
   c. **Lateral**
      (1) Skeletal muscles of the limbs
      (2) Sternum, clavicle, scapula, pelvis, and the bones of the limbs
      (3) Serous membranes of body cavities
      (4) Lamina propria, muscularis mucosae, submucosa, muscularis externae, and adventitia of the gastrointestinal tract
      (5) Blood cells, microglia, Kupffer cells
      (6) Cardiovascular system
      (7) Lymphatic system
      (8) Spleen
      (9) Adrenal cortex
      (10) Laryngeal cartilages

3. **Endoderm**
   a. Epithelial lining of the auditory tube and middle ear cavity
   b. Epithelial lining of the posterior one-third of the tongue, floor of the mouth, palatoglossal and palatopharyngeal folds, soft palate, crypts of the palatine tonsil, and sublingual and submandibular glands and ducts
   c. Principal and oxyphil cells of the parathyroid glands
   d. Epithelial reticular cells and thymic corpuscles
   e. Thyroid follicular cells
   f. Epithelial lining and glands of the trachea, bronchi, and lungs
   g. Epithelial lining of the gastrointestinal tract
   h. Hepatocytes and epithelial lining of the biliary tree
   i. Acinar cells, islet cells, and the epithelial lining of the pancreatic ducts
   j. Epithelial lining of the urinary bladder
   k. Epithelial lining of the vagina
   l. Epithelial lining of the female urethra and most of the male urethra

5

Placenta, Amniotic Fluid, and Umbilical Cord

I. PLACENTA

A. Formation (Figure 5-1)

1. Components

a. The **maternal component** of the placenta consists of the **decidua basalis,** a portion of the endometrium.

b. The **fetal component** of the placenta consists of the **villous chorion.**

2. Clinical correlations

a. **Placental previa** occurs when the placenta attaches in the lower part of the uterus and covers the internal os. **Uterine blood vessels** may rupture during the advanced stages of pregnancy, causing a potentially fatal hemorrhage in the mother and placing the fetus in jeopardy as a result of the compromised blood supply.

b. **Twinning** (Figure 5-2)

(1) **Monozygotic (identical) twins** develop from one zygote.

(a) In **65%** of cases, the fetuses have **one** placenta, **one** chorion, and **two** amniotic sacs.

(b) In the **remaining 35% of cases,** the fetuses have **two** placentas (separate or fused), **two** chorions, and **two** amniotic sacs.

(2) **Dizygotic (fraternal) twins** develop from two zygotes. The fetuses have **two** placentas, **two** chorions, and **two** amniotic sacs.

B. Placental membrane (fetal–maternal barrier)

1. Layers

a. In **early pregnancy,** the placental membrane consists of the **syncytiotrophoblast, cytotrophoblast, connective tissue,** and the **endothelium of the fetal capillaries.**

b. In **late pregnancy,** the cytotrophoblast cells degenerate and the connective tissue is displaced by the growth of fetal blood vessels, leaving the **syncytiotrophoblast** and the **fetal capillary endothelium.**

2. **Function.** The placental membrane separates maternal blood from fetal blood. Some substances (both beneficial and harmful) cross the placental membrane freely, whereas it is impermeable to others (Tables 5-1, 5-2, and 5-3).

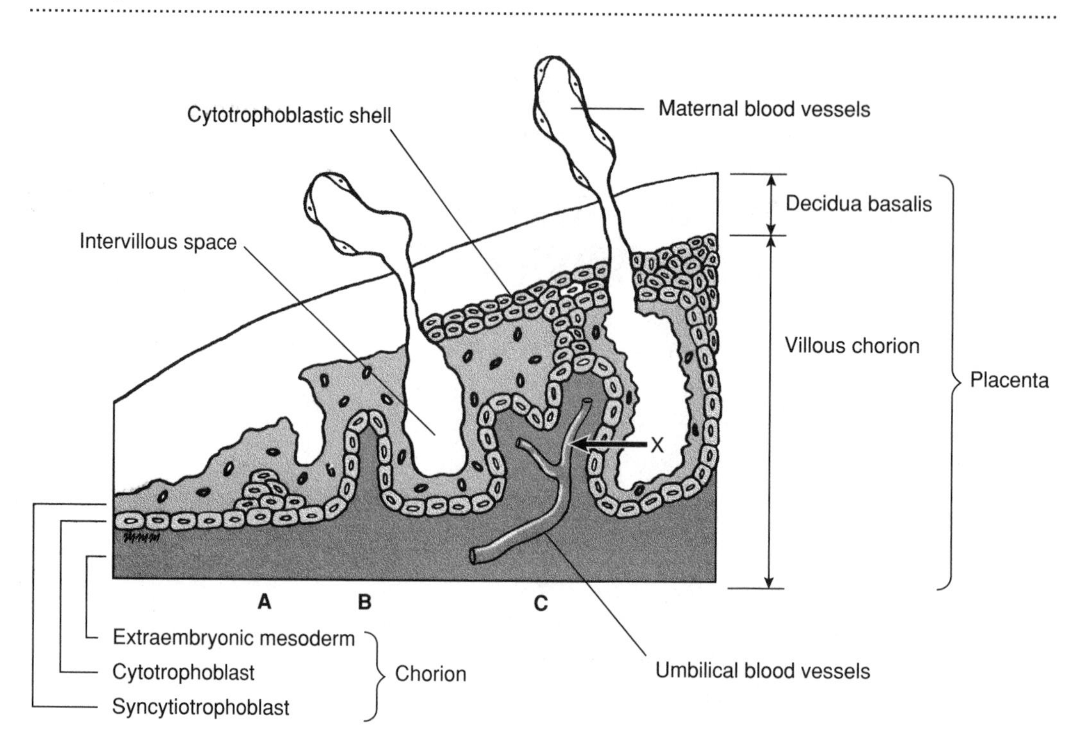

Figure 5-1. Placenta formation. (*A*) Primary chorionic villi form during week 2 when the cytotrophoblast cells invade the syncytiotrophoblast. (*B*) Secondary chorionic villi form during week 3 when extraembryonic mesoderm evaginates into the cytotrophoblast. (*C*) Tertiary chorionic villi form when the umbilical blood vessels develop in the extraembryonic mesoderm. Tertiary chorionic villi collectively comprise the villous chorion. The X indicates maternal blood within the intervillous space, and the *arrow* indicates the layers that substances must pass through during fetal–maternal blood exchange. In late pregnancy, the placental membrane (fetal–maternal barrier) is reduced to two layers.

3. **Clinical correlations. Erythroblastosis fetalis** occurs when fetal erythrocytes are Rh-positive but the maternal erythrocytes are Rh-negative. When the fetal erythrocytes cross the placental membrane and enter the maternal circulation, the mother's body forms anti-Rh antibodies that cross the placental membrane and destroy the erythrocytes of the fetus.

II. AMNIOTIC FLUID

A. Production. Amniotic fluid is produced by **dialysis of maternal** and **fetal blood** through blood vessels in the placenta and by **excretion of fetal urine** into the amniotic sac.

B. Resorption. After being swallowed by the fetus, the amniotic fluid is absorbed into the fetal bloodstream. Excess amniotic fluid is removed by the placenta and passed into the maternal blood.

C. Clinical correlations

1. **Oligohydramnios** occurs when amniotic fluid is deficient (less than 400 ml in late pregnancy). Oligohydramnios may be associated with the inability of the fetus to excrete urine into the amniotic sac as a result of **renal agenesis.**

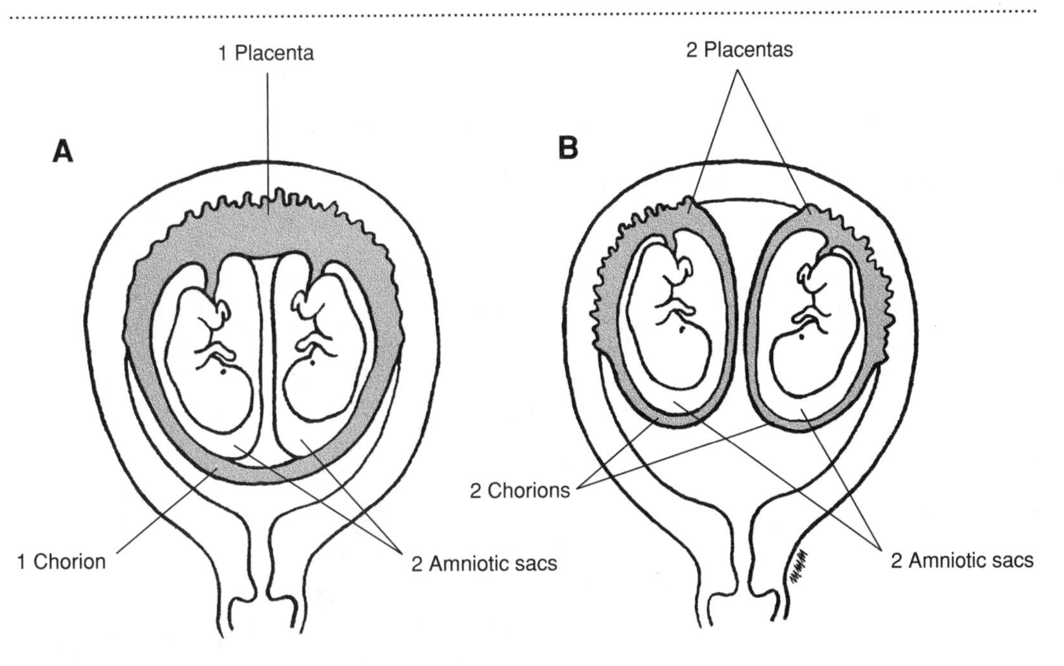

Figure 5-2. Arrangement of the placenta, chorion, and amniotic sac of (*A*) 65% of monozygotic twins and (*B*) 35% of monozygotic twins and all dizygotic twins.

Table 5-1
Beneficial Substances That Cross the Placental Membrane

Oxygen, carbon dioxide
Nutrients (e.g., glucose, amino acids, free fatty acids, vitamins)
Electrolytes (e.g., sodium, potassium, chloride, calcium, phosphate)
Water
Fetal waste products (e.g., carbon dioxide, urea, uric acid, bilirubin)
Fetal and maternal erythrocytes
Maternal serum proteins
Steroid hormones
Immunoglobulin G (IgG)

Table 5-2
Harmful Substances That Cross the Placental Membrane

Carbon monoxide
Viruses (e.g., HIV, cytomegalovirus, rubella, Coxsackie, variola, varicella, measles, poliomyelitis)
Treponema palladium, Toxoplasma gondii
Drugs (e.g., cocaine, alcohol, caffeine, nicotine, warfarin, trimethadione, phenytoin, cancer chemotherapeutic agents, anesthetics, sedatives, analgesics)
Anti-Rh antibodies

HIV = human immunodeficiency virus.

Table 5-3
Substances That Do Not Cross the Placental Membrane

Maternaliy-derived cholesterol, triglycerides, and phospholipids
Protein hormones (e.g., insulin)
Immunoglobulin M (IgM)
Succinylcholine, curare, heparin, drugs similar to amino acids (e.g., methyldopa)
Most bacteria

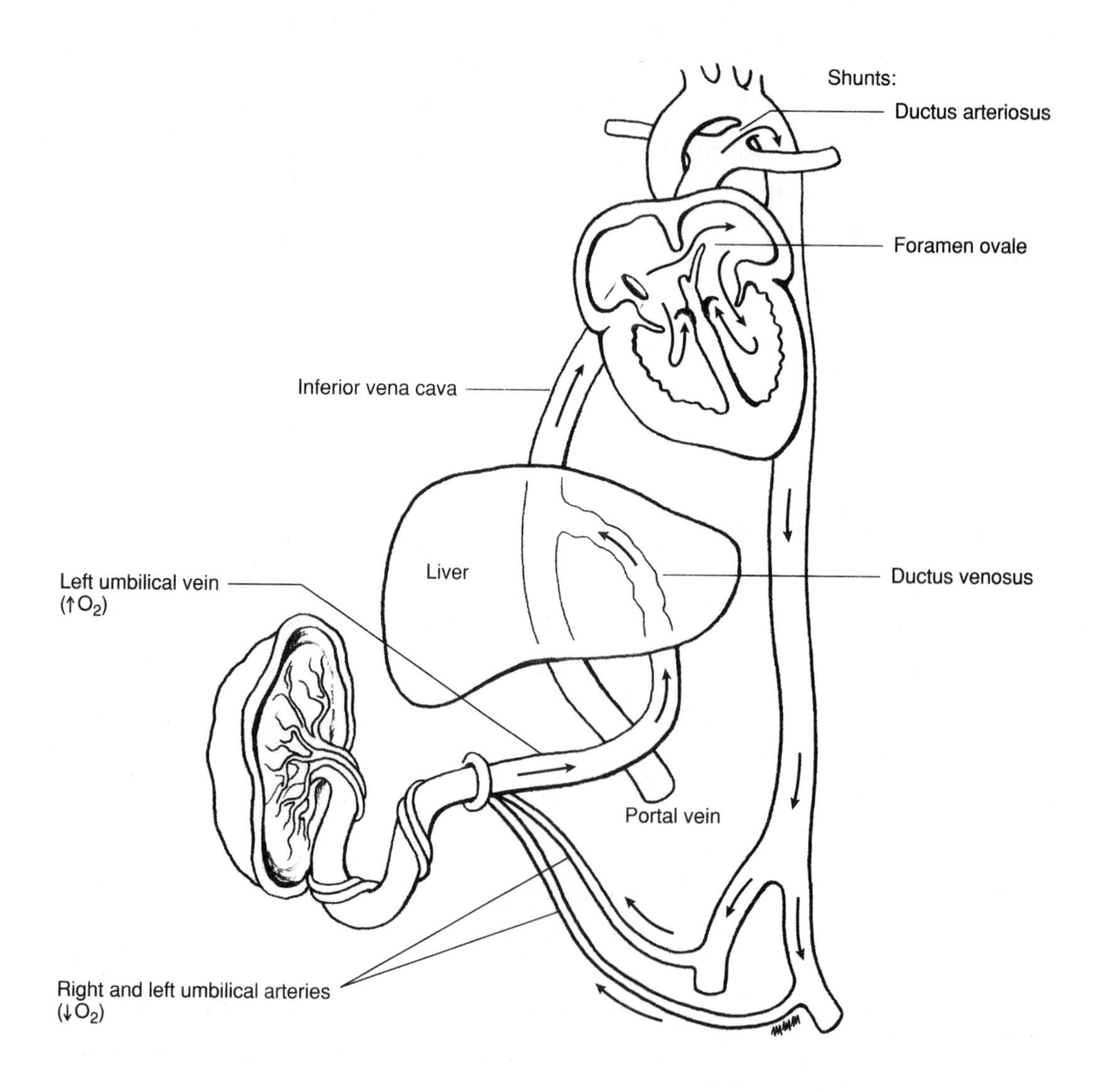

Figure 5-3. Fetal circulation.

2. **Polyhydramnios** occurs when amniotic fluid levels are high (more than 2000 ml in late pregnancy). This condition may be associated with **anencephaly** or **esophageal atresia.**

3. **α-Fetoprotein (AFP) assay** is used to diagnose neural tube defects (e.g., spina bifida, anencephaly). AFP is found in amniotic fluid and maternal serum.

III. UMBILICAL CORD.

The definitive umbilical cord contains the right and left umbilical arteries, the left umbilical vein, and mucous connective tissue.

A. The **umbilical arteries** carry **deoxygenated blood from the fetus to the placenta.**

B. The left **umbilical vein** carries **oxygenated blood from the placenta to the fetus.**

C. Clinical correlations

1. **Vasa previa** occurs when the **umbilical blood vessels** cross the internal os of the uterus, predisposing them to rupture. Rupture of the umbilical blood vessels during pregnancy, labor, or delivery can cause the fetus to bleed to death.

2. **Presence of one umbilical artery,** as opposed to two, may suggest the presence of cardiovascular abnormalities.

IV. ANGIOGENESIS AND HEMATOPOIESIS.

Mesoderm differentiates into **angioblasts,** which form **angiogenic cell clusters.**

A. Angiogenesis. Angioblasts around the periphery of the angiogenic cell clusters give rise to the **endothelium of blood vessels.** Angiogenesis occurs initially in **extraembryonic visceral mesoderm** (located around the yolk sac), and later in the mesoderm of the fetus.

B. Hematopoiesis. Angioblasts within the center of the angiogenic cell clusters give rise to **blood cells.** Hematopoiesis occurs initially in the **extraembryonic visceral mesoderm** (located around the yolk sac) and later in the fetal liver, spleen, thymus, and bone marrow.

V. FETAL CIRCULATION

FETAL CIRCULATION involves three shunts: the **ductus venosus, ductus arteriosus,** and **foramen ovale** (Figure 5-3). A number of changes occur in the neonatal circulation when placental blood flow ceases and lung respiration begins. Table 5-4 summarizes the remnants that result from closure of the fetal structures.

Table 5-4
Remnants Created by Closure of Fetal Circulatory Structures

Fetal Structure	Adult Remnant
Right and left umbilical arteries	Medial umbilical ligaments
Left umbilical vein	Ligamentum teres
Ductus venosus	Ligamentum venosum
Ductus arteriosus	Ligamentum arteriosum
Foramen ovale	Fossa ovale

6

Cardiovascular System

I. DEVELOPMENT OF THE HEART

A. Primitive heart tube. A pair of **endocardial heart tubes** (mesodermal in origin) form within the cardiogenic region.

1. As lateral folding occurs, the endocardial heart tubes fuse to form the **primitive heart tube,** which develops into the **endocardium.**

2. Mesoderm surrounding the primitive heart tube develops into the **myocardium** and **epicardium.**

3. The primitive heart tube forms **five dilatations** (Figure 6-1), the fates of which are detailed in Table 6-1.

B. Aorticopulmonary (AP) septum. The AP septum divides the truncus arteriosus into the **aorta** and **pulmonary trunk.**

1. Formation (Figure 6-2). Neural crest cells migrate into the **truncal** and **bulbar ridges,** which grow and twist around each other in a spiral fashion, fusing to form the AP septum.

2. Clinical correlations

a. Transposition of the great vessels occurs when the AP septum fails to develop in a spiral fashion, causing the aorta to open into the right ventricle and the pulmonary trunk to open into the left ventricle. The resultant **right-to-left shunting** of blood leads to **cyanosis.**

b. Tetralogy of Fallot (Figure 6-3) occurs when the AP septum fails to align properly with the atrioventricular septum and results in **pulmonary stenosis, overriding aorta, interventricular septal defect,** and **right ventricular hypertrophy.** Tetralogy of Fallot is characterized by **right-to-left shunting** of blood and **cyanosis.**

C. Atrioventricular (AV) septum. The AV septum partitions the AV canal into the **right AV canal** and **left AV canal.** The dorsal and ventral **AV cushions** fuse to form the AV septum (Figure 6-4).

D. Atrial septum

1. Formation (Figure 6-5)

a. The **septum primum** grows toward the AV septum.

b. The **foramen primum** is located between the edge of the septum primum and the AV septum; it is obliterated when the septum primum fuses with the AV septum.

c. The **foramen secundum** forms in the center of the septum primum.

d. The **septum secundum** forms to the right of the septum primum and fuses (after birth) with the septum primum to form the atrial septum.

e. The **foramen ovale** is the opening between the upper and lower parts of the septum secundum.

(1) During fetal life, blood is shunted from the right atrium to the left atrium via the foramen ovale.

(2) **Closure of the foramen ovale** normally takes place soon after birth and

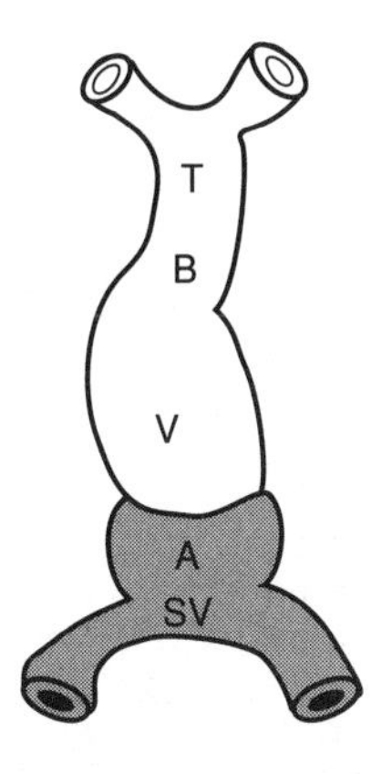

Figure 6-1. The five dilatations of the heart tube. *A* = primitive atrium; *B* = bulbus cordis, *SV* = sinus venosus; *T* = truncus arteriosus; *V* = primitive ventricle; *white area* = arterial portion; *shaded area* = venous portion. (Modified with permission from Fix JD and Dudek RW: *BRS Embryology*, Baltimore, Williams & Wilkins, 1995, p 40.)

Table 6-1
Structures Derived From the Embryonic Dilatations of the Primitive Heart Tube

Embryonic Dilatation	Adult Structure
Truncus arteriosus	Aorta
	Pulmonary trunk
Bulbus cordis	Smooth part of right ventricle **(conus arteriosus)**
	Smooth part of left ventricle **(aortic vestibule)**
Primitive ventricle	Trabeculated part of right ventricle
	Trabeculated part of left ventricle
Primitive atrium	Trabeculated part of right atrium
	Trabeculated part of left atrium
Sinus venosus	Smooth part of right atrium **(sinus venarum)***
	Coronary sinus
	Oblique vein of left atrium

a) The smooth part of the left atrium is formed by incorporation of parts of the **pulmonary veins** into the atrial wall.
b) The junction of the trabeculated and smooth parts of the right atrium is called the **crista terminalis.**

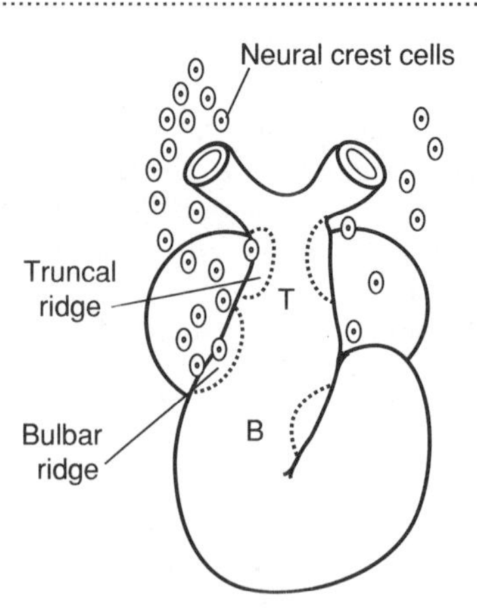

Figure 6-2. Formation of the aorticopulmonary (AP) septum. *B* = bulbus cordis; *T* = truncus arteriosus. (Modified with permission from Fix JD and Dudek RW: *BRS Embryology*, Baltimore, Williams & Wilkins, 1995, p 41.)

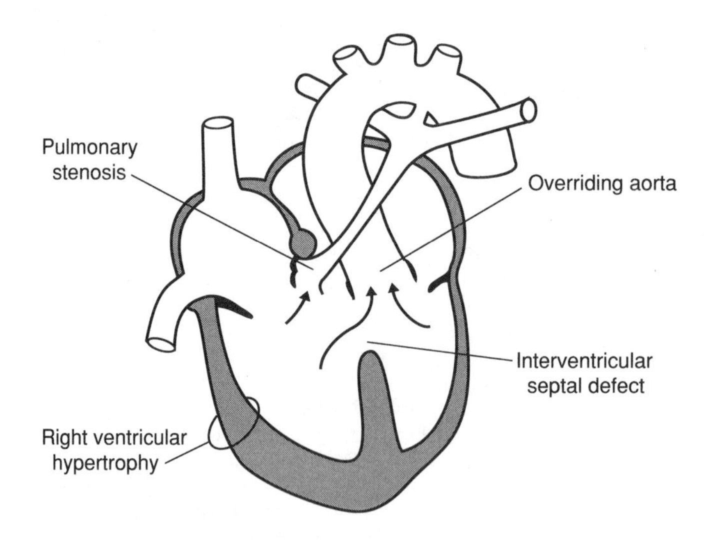

Figure 6-3. The four main defects associated with tetralogy of Fallot. (Modified with permission from Fix JD and Dudek RW: *BRS Embryology*, Baltimore, Williams & Wilkins, 1995, p 44.)

is facilitated by the **increased left atrial pressure** that results from changes in the pulmonary circulation.

2. Clinical correlations. Foramen secundum defect is caused by excessive resorption of the septum primum or septum secundum and results in a **patent foramen ovale.**

E. Interventricular (IV) septum (Figure 6-6)

1. Formation

a. The **muscular IV septum** develops in the floor of the ventricle and grows toward the AV septum, stopping short to create the **IV foramen.**

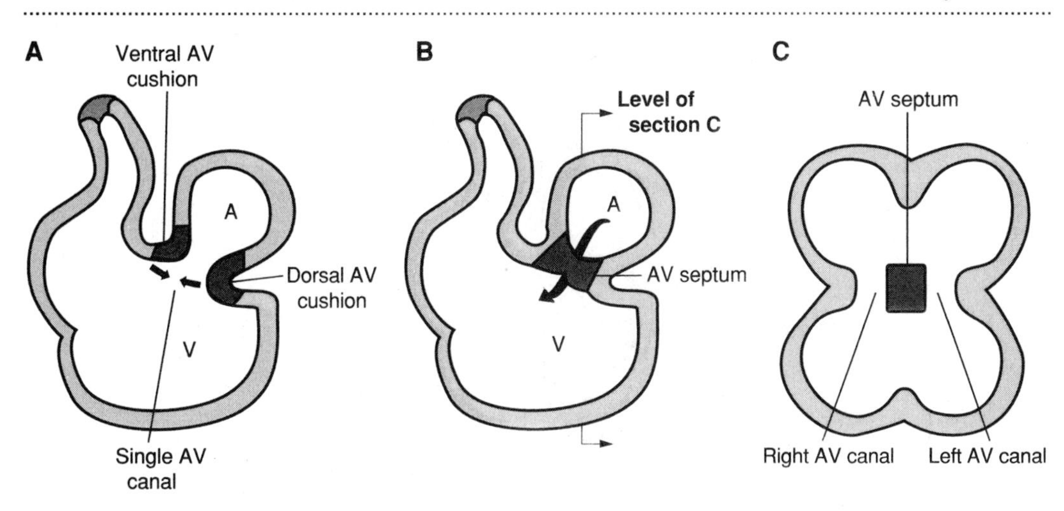

Figure 6-4. Formation of the atrioventricular (AV) septum. *A* = atrium; *V* = ventricle. (Reprinted with permission from Fix JD and Dudek RW: *BRS Embryology*, Baltimore, Williams & Wilkins, 1995, p 42.)

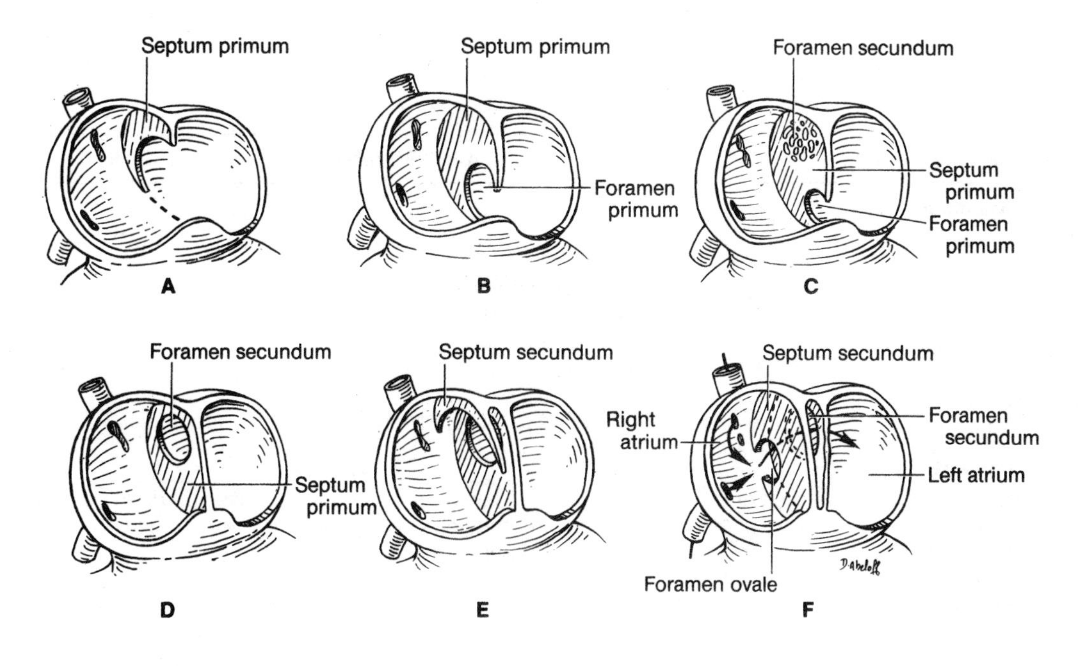

Figure 6-5. Formation of the atrial septum. The *arrows* in *F* indicate the direction of blood flow across the fully developed septum, from the right atrium to the left atrium. (Modified with permission from Johnson KE: *NMS Human Developmental Anatomy*, Baltimore, Williams & Wilkins, 1988, p 149.)

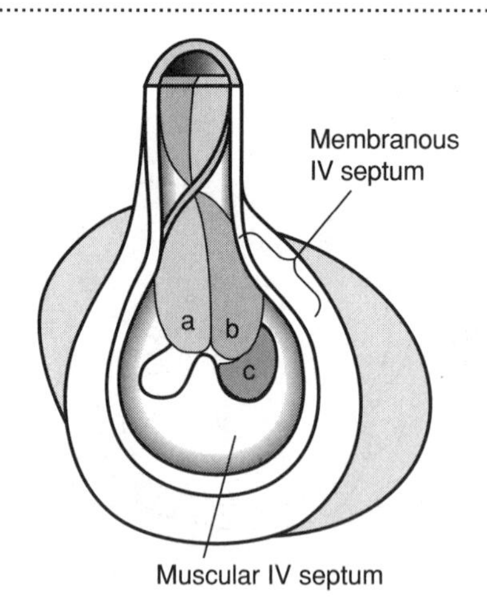

Figure 6-6. The interventricular (IV) septum. *a* = right bulbar ridge; *b* = left bulbar ridge; *c* = atrioventricular (AV) septum. (Modified with permission from Fix JD and Dudek RW: *BRS Embryology*, Baltimore, Williams & Wilkins, 1995, p 43.)

Table 6-2
Development of the Arterial System

Embryonic Structure	Adult Structure
Aortic arches	
1	*
2	*
3	Right and left common carotid arteries Proximal right and left internal carotid arteries
4	Proximal right subclavian artery Part of the aortic arch
5	**
6	Proximal right and left pulmonary arteries; the ductus arteriosus Ductus arteriosus
Dorsal aorta	
Posterolateral branches	Intercostal, lumbar, lateral sacral, median sacral, vertebral, deep cervical, ascending cervical, internal thoracic, superior and inferior epigastric arteries, and arteries to the upper and lower extremities
Lateral branches	Renal, suprarenal, and gonadal arteries
Ventral branches	
Vitelline arteries	Celiac, superior mesenteric, and inferior mesenteric arteries
Umbilical arteries	Part of the internal iliac arteries, superior vesical arteries, and medial umbilical ligaments

* Minimal contribution in the adult.
** Aortic arch 5 degenerates in humans.

Table 6-3
Development of the Venous System

Embryonic Structure	Adult Structure
Vitelline veins	
Right	Hepatic veins and sinusoids, ductus venosus, part of the IVC, and portal, superior mesenteric, inferior mesenteric, splenic veins
Left	Hepatic veins and sinusoids, ductus venosus
Umbilical veins	
Right	Degenerates
Left	Ligamentum teres
Cardinal veins	
Anterior cardinal	Internal jugular veins, SVC
Posterior cardinal	Part of IVC, common iliac veins
Subcardinal	Part of IVC, renal veins, gonadal veins
Supracardinal	Part of IVC, intercostal veins, hemiazygos vein, azygos vein

IVC = inferior vena cava; SVC = superior vena cava.

b. The **membranous IV septum** forms following fusion of the **right** and **left bulbar ridges** and the **AV septum.** The membranous IV septum closes the IV foramen.

2. **Clinical correlations. Membranous ventricular septal defect (VSD)** is caused by failure of the membranous IV septum to develop, resulting in **left-to-right shunting** of blood through the IV foramen. Left-to-right shunting of blood increases blood flow to the lungs, resulting in **pulmonary hypertension.** Patients with left-to-right shunting complain of excessive fatigue on exertion.

II. DEVELOPMENT OF THE ARTERIAL SYSTEM

A. Formation. The arterial system develops from the **aortic arches** and branches of the **dorsal aorta** (Table 6-2).

B. Clinical correlations

1. Postductal coarctation of the aorta occurs when the aorta is abnormally constricted just inferior to the ductus arteriosus. This condition is clinically associated with increased blood pressure in the upper extremities, lack of a femoral arterial pulse, and a high risk of cerebral hemorrhage and bacterial endocarditis.

2. Patent ductus arteriosus (PDA) occurs when the ductus arteriosus fails to close after birth.

a. Normally, the ductus arteriosus closes within a few hours via smooth muscle contraction, forming the **ligamentum arteriosum.**

b. PDA is common in **premature infants** and those born to mothers infected with the **rubella** virus during the course of their pregnancy.

III. DEVELOPMENT OF THE VENOUS SYSTEM.

The venous system develops from the **vitelline, umbilical,** and **cardinal** veins (Table 6-3).

7

Gastrointestinal System

I. PRIMITIVE GUT TUBE. The primitive gut tube is divided into the **foregut, midgut,** and **hindgut** (Figure 7-1).

A. Formation. The primitive gut tube is formed by the incorporation of the yolk sac into the embryo during craniocaudal and lateral folding of the embryo.

1. The epithelial lining and glands of the gut tube mucosa are derived from **endoderm,** whereas the lamina propria, muscularis mucosae, submucosa, muscularis externa, and adventitia or serosa are derived from **mesoderm.**

2. The epithelial lining of the gut tube proliferates rapidly and obliterates the lumen. **Canalization** then occurs.

B. Foregut derivatives are supplied by the **celiac artery.**

1. Esophagus

a. Formation. The **tracheoesophageal septum** divides the foregut into the esophagus and trachea.

b. Clinical correlations. Esophageal atresia occurs when the esophagus ends as a blind tube as a result of malformation of the tracheoesophageal septum.

2. Stomach

a. Formation. The stomach develops from a fusiform dilatation that forms in the foregut during week 4. The primitive stomach rotates 90° clockwise during its formation, causing the formation of the **lesser peritoneal sac.**

b. Clinical correlations. Hypertrophic pyloric stenosis occurs when the muscularis externa hypertrophies, narrowing the pyloric lumen. This condition is associated with projectile vomiting and a small, palpable mass at the right costal margin.

3. Liver

a. An endodermal outgrowth of the foregut (the **hepatic diverticulum**) forms in the surrounding mesoderm (i.e., the **septum transversum**). (The septum transversum also plays a role in the formation of the **diaphragm,** which explains the proximity of the liver and diaphragm in adults.)

b. The hepatic diverticulum sends **hepatic cell cords** into the septum transversum. The hepatic cell cords surround the vitelline veins, which form **hepatic sinusoids.**

4. Gallbladder and bile ducts

a. Formation. The connection between the hepatic diverticulum and foregut narrows to form the bile duct. Later, an outgrowth from the bile duct gives rise to the gallbladder and cystic duct.

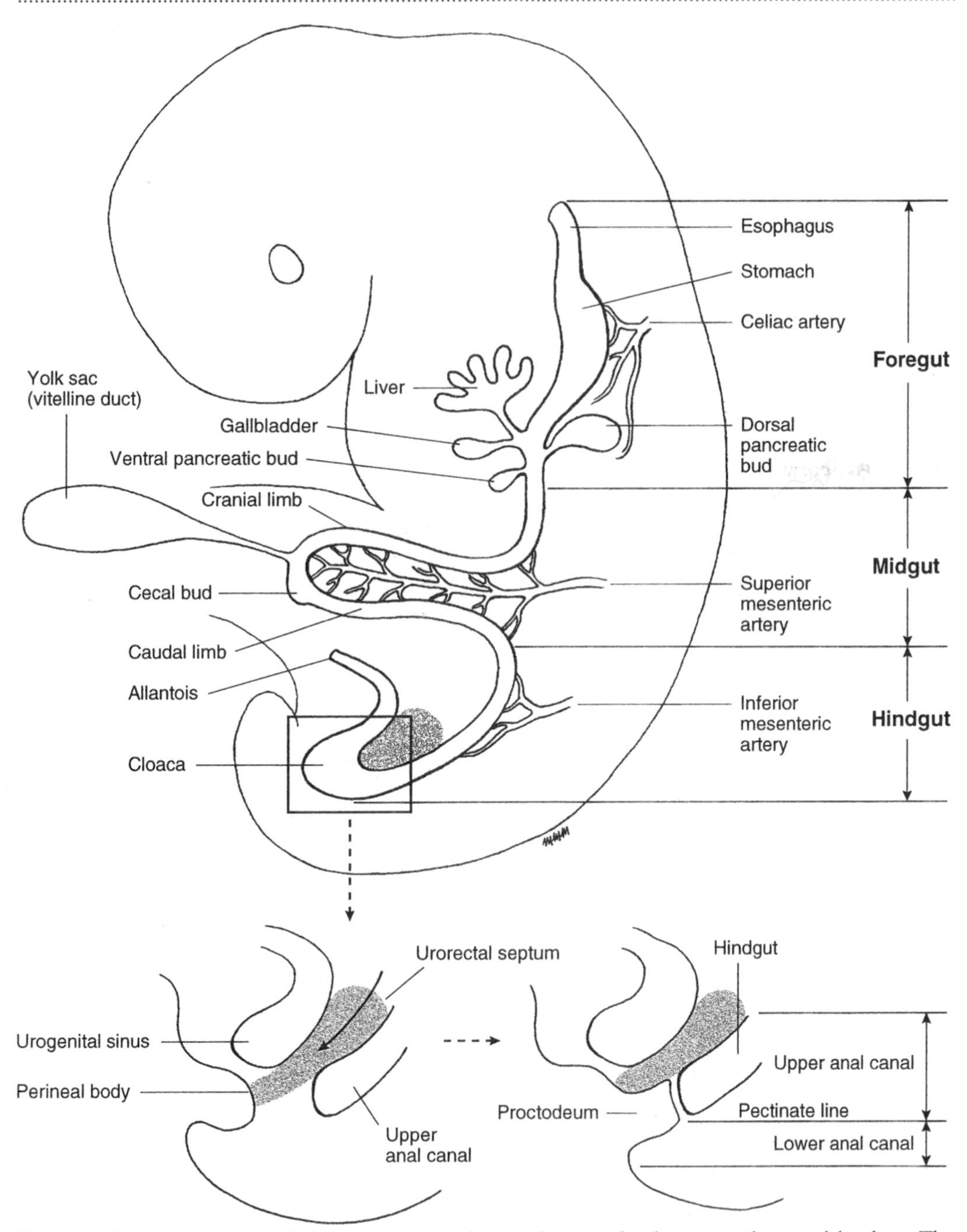

Figure 7-1. Development of the gastrointestinal tract showing the foregut, midgut, and hindgut. The urorectal septum partitions the cloaca into the upper anal canal and the urogenital sinus (*boxed area*). The *curved arrow* indicates the direction of growth of the urorectal septum toward the body surface, where it fuses at the perineal body. The ectoderm invaginates, forming the **proctodeum,** and meets the upper anal canal at the **pectinate line** to form the lower anal canal.

b. Clinical considerations. Extrahepatic biliary atresia occurs when incomplete canalization leads to occlusion of the biliary duct lumina. This condition is associated with jaundice, pale feces, and dark urine.

5. Pancreas

a. Formation

(1) The **ventral pancreatic bud** forms the uncinate process and part of the head of the pancreas.

(2) The **dorsal pancreatic bud** forms the remaining part of the head, body, and tail of the pancreas.

b. Clinical correlations. Annular pancreas occurs when the ventral and dorsal pancreatic buds form a ring around the duodenum, thereby obstructing it.

6. Upper duodenum. The cranial portion of the duodenum develops from the caudal portion of the foregut. The junction of the foregut and midgut is just **distal to the opening of the common bile duct.**

C. Midgut derivatives are supplied by the **superior mesenteric artery.**

1. Lower duodenum

a. Formation. The lower duodenum originates from the cranial portion of the midgut.

b. Clinical correlations. Duodenal atresia occurs when the lumen of the duodenum is occluded as a result of failed recanalization. This condition is associated with polyhydramnios, bile-containing vomitus, and distention of the stomach.

2. Jejunum, ileum, cecum, appendix, ascending colon, and the proximal two-thirds of the transverse colon

a. Formation

(1) The **midgut loop** herniates through the umbilicus during a process known as **physiologic umbilical herniation.**

(a) The **cranial limb** of the midgut loop forms the **jejunum** and **cranial portion of the ileum.**

(b) The **caudal limb** forms the **cecum, appendix, caudal portion of the ileum, ascending colon,** and **proximal two-thirds of the transverse colon.**

(2) The midgut loop rotates **270° counterclockwise** around the superior mesenteric artery as it returns to the abdominal cavity, reducing the physiologic herniation.

b. Clinical correlations

(1) Omphalocoele occurs when the midgut loop fails to return to the abdominal cavity. In the newborn, a light gray, shiny sac is visible at the base of the umbilical cord.

(2) Ileal (Meckel's) diverticulum occurs when a remnant of the yolk sac (or vitelline duct) persists, forming an opening between the lumen of the ileum and the umbilicus. This condition is associated with drainage of meconium from the umbilicus.

(3) Malrotation of the midgut occurs when the midgut undergoes only partial rotation, resulting in abnormal positioning of the abdominal viscera. This condition may be associated with **volvulus.**

(4) Intestinal atresia or stenosis occurs as a result of failed recanalization.

D. Hindgut derivatives (i.e., the **distal third of the transverse colon, descending colon, sigmoid colon,** and **upper anal canal**) are supplied by the **inferior mesenteric artery.**

Table 7-1
Derivation of Adult Mesenteries

Embryonic Mesentery	Adult Mesentery
Ventral	Lesser omentum (hepatoduodenal and hepatogastric ligaments), falciform ligament, coronary ligament, triangular ligament
Dorsal	Greater omentum (gastrorenal, gastrosplenic, gastrocolic, and splenorenal ligaments), mesentery of small intestine, mesoappendix, transverse mesocolon, sigmoid mesocolon

1. Formation
   a. The **cranial end** of the hindgut forms the **distal third of the transverse colon,** the **descending colon,** and the **sigmoid colon.**
   b. The **terminal end** of the hindgut (the **cloaca**) is partitioned by the **urorectal septum** into the **upper anal canal** and the **urogenital sinus** (see Figure 7-1).
2. Clinical correlations
   a. **Colonic aganglionosis (Hirschsprung's disease)** results from failure of **neural crest cells** to form the myenteric plexus in the sigmoid colon and rectum. This condition is associated with loss of peristalsis, fecal retention, and abdominal distention.
   b. **Anorectal agenesis** occurs when the rectum ends as a blind sac **above** the puborectalis muscle because of abnormal formation of the urorectal septum. This condition may be accompanied by a **rectovesical fistula, rectourethral fistula,** or **rectovaginal fistula.**
   c. **Anal agenesis** occurs when the anal canal ends as a blind sac **below** the puborectalis muscle as a result of abnormal formation of the urorectal septum. This condition may be accompanied by a **rectovesical fistula, rectourethral fistula,** or **rectovaginal fistula.**

II. PROCTODEUM. The **lower anal canal** develops from an invagination of surface **ectoderm** called the **proctodeum** (see Figure 7-1).

III. MESENTERIES. The primitive gut tube is suspended within the peritoneal cavity of the embryo by the **ventral** and **dorsal mesenteries,** from which all adult mesenteries are derived (Table 7-1).

8

Respiratory System

I. UPPER RESPIRATORY SYSTEM. The upper respiratory system is discussed in Chapter 9.

II. LOWER RESPIRATORY SYSTEM. The **trachea, bronchi,** and **lungs** comprise the lower respiratory system.

A. Formation (Figure 8-1)

1. The **laryngotracheal diverticulum** forms in the ventral wall of the foregut.
2. The **tracheoesophageal septum** divides the foregut into the esophagus and trachea.
3. The distal end of the laryngotracheal diverticulum enlarges to form the **lung bud.**

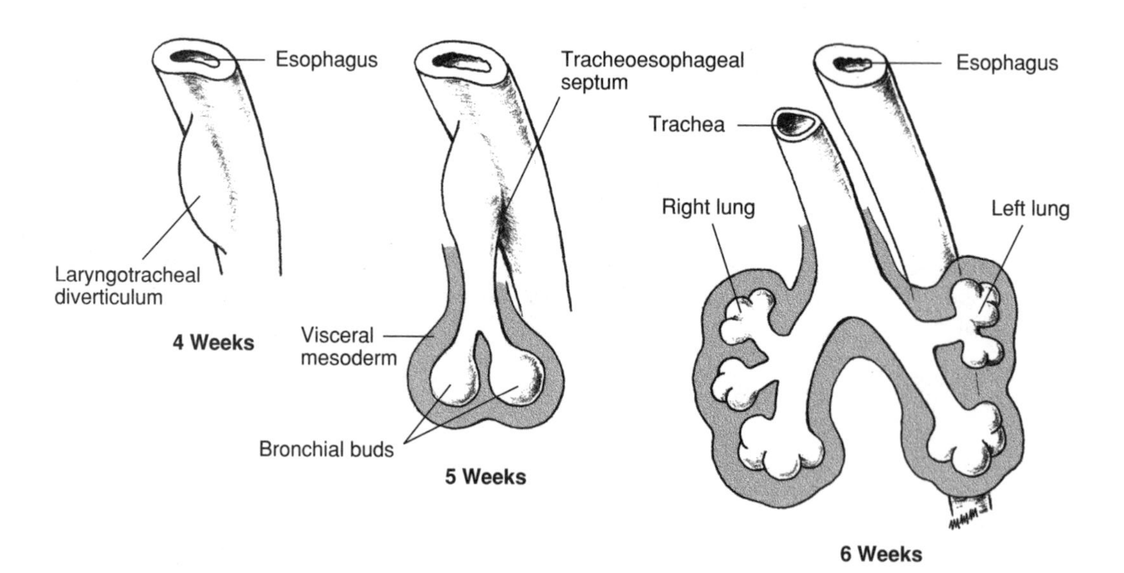

Figure 8-1. Development of the respiratory system. (Modified with permission from Johnson KE: *NMS Human Developmental Anatomy*, Baltimore, Williams & Wilkins, 1988, p 228.)

Table 8-1
Stages of Lung Development

Name	Time Period	Characteristics
Glandular	Weeks 5–17	Respiration is not possible Premature fetuses cannot survive
Canalicular	Weeks 13–25	Respiratory bronchioles and terminal sacs form Vascularization increases Premature fetuses born before week 20 rarely survive
Terminal Sac	Weeks 24–birth	Type I and Type II pneumocytes are present Respiration is possible Premature fetuses born between weeks 25 and 28 can survive with intensive care
Alveolar	Birth–year 8	Respiratory bronchioles, terminal sacs, alveolar ducts, and alveoli increase in number

4. The lung bud divides into two **bronchial buds,** which branch into the **primary, secondary,** and **tertiary bronchi.** The tertiary bronchi are related to the **bronchopulmonary segments** in the adult lung.

B. Stages of development. The lungs undergo four stages of development (Table 8-1).

C. Clinical correlations

1. **Tracheoesophageal fistula.** A tracheoesophageal fistula is an abnormal communication between the trachea and esophagus caused by a malformation of the tracheoesophageal septum. This condition results in gagging and cyanosis after feeding, abdominal distention after crying, and reflux of the gastric contents into the lungs.

2. **Respiratory distress syndrome** is caused by a deficiency of surfactant. This condition is most common in premature infants, those born to diabetic mothers, and those experiencing prolonged intrauterine asphyxia. Treatment with thyroxine and cortisol can increase the production of surfactant.

3. **Pulmonary hypoplasia** occurs when lung development is stunted. **Congenital diaphragmatic hernia** (herniation of the abdominal contents into the thorax leads to compression of the lung) and **bilateral renal agenesis** (oligohydramnios increases the pressure on the fetal thorax) are predisposing factors.

9

Head and Neck

I. PHARYNGEAL APPARATUS (Figure 9-1). The pharyngeal apparatus consists of the pharyngeal **arches, pouches, grooves,** and **membranes.**

A. The **pharyngeal arches (1,2,3,4,6)*** are comprised of **mesoderm** and **neural crest cells.** Each arch has a **cranial nerve** associated with it. Table 9-1 summarizes the adult derivatives of the pharyngeal arches.

B. The **pharyngeal pouches (1,2,3,4)** are evaginations of the endoderm-lined foregut. Table 9-2 summarizes the adult derivatives of the pharyngeal pouches.

C. The **pharyngeal grooves (1,2,3,4)** are invaginations of surface ectoderm.

1. Pharyngeal groove 1 gives rise to the **epithelial lining of the external auditory meatus.**

2. All of the other grooves are obliterated.

D. The **pharyngeal membranes (1,2,3,4)** are located at the junction of each pharyngeal groove and pouch.

1. Pharyngeal membrane 1 gives rise to the **tympanic membrane.**

2. All of the other membranes are obliterated.

II. THYROID GLAND. The thyroid gland develops from the **thyroid diverticulum,** which forms in the floor of the foregut. The thyroid diverticulum migrates caudally to its adult anatomical position but remains connected to the foregut via the **thyroglossal duct,** which is later obliterated. The former site of the thyroglossal duct is indicated in the adult by the **foramen cecum** (see Figure 9-1B).

III. TONGUE

A. The **anterior two-thirds** of the tongue form from the **median tongue bud** and two **distal tongue buds** associated with pharyngeal arch 1.

* Pharyngeal arch 5 degenerates in humans.

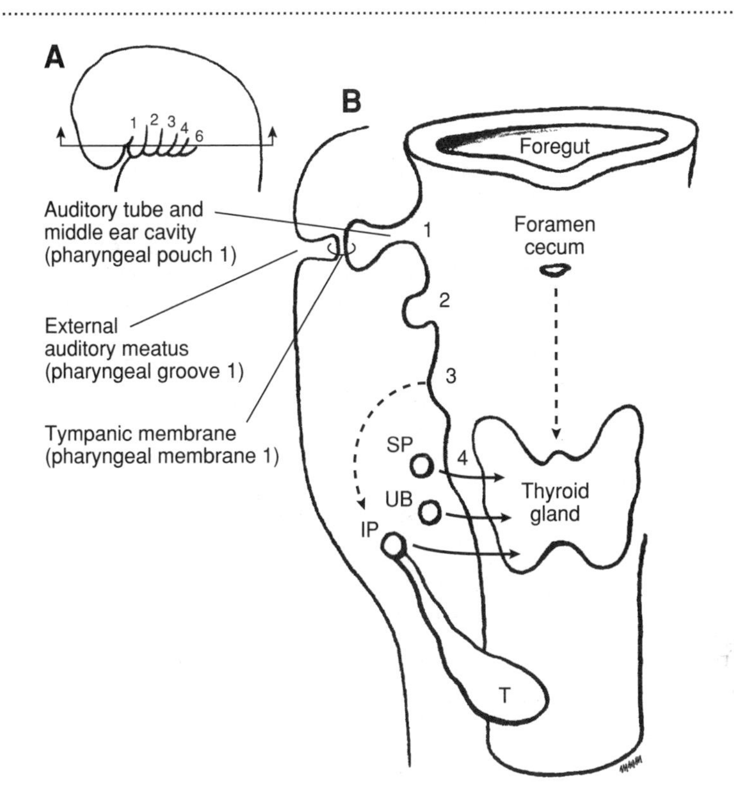

Figure 9-1. (*A*) Overview of the pharyngeal apparatus. (*B*) Migration of the superior and inferior parathyroid glands (*SP*, *IP*), thymus (*T*), ultimobranchial body (*UB*), and thyroid gland. The foramen cecum evaginates to form the thyroid diverticulum, which migrates along the midline (*dotted arrow*). In addition, pharyngeal pouch 1, pharyngeal membrane 1, and pharyngeal groove 1 are shown. These structures give rise to structures of the adult ear. *2* = pharyngeal pouch 2; *3* = pharyngeal pouch 3; *4* = pharyngeal pouch 4.

1. General sensation is carried by the **lingual branch of CN V.**
2. Taste sensation is carried by the **chorda tympani branch of CN VII.**

B. The **posterior one-third** of the tongue forms predominately from the **hypobranchial eminence** associated with pharyngeal arches 3 and 4. General sensation and taste are carried by **CN IX.**

C. The **intrinsic muscles** and the **styloglossus, hyoglossus,** and **genioglossus** (extrinsic) **muscles** are derived from myoblasts that migrate to the tongue region from **occipital somites.** Motor innervation is supplied by **CN XII,** except for that of the palatoglossus muscle, which is innervated by CN X.

IV. PALATE (Figure 9-2)

A. The **intermaxillary segment** forms when the two medial nasal prominences fuse together at the midline. The intermaxillary segment gives rise to the **philtrum of the lip, four incisor teeth,** and **primary palate** of the adult.

Table 9-1
Adult Derivatives of the Pharyngeal Arches

Arch	Nerve	Adult Derivatives: Mesoderm	Adult Derivatives: Neural Crest Cells
1	CN V	Muscles of mastication, mylohyoid muscle, tensor veli palatini muscle, tensor tympani muscle, anterior belly of the digastric muscle	Maxilla, zygomatic bone, temporal bone, palatine bone, vomer, mandible, incus, malleus, sphenomandibular ligament
2	CN VII	Muscles of facial expression, posterior belly of the digastric muscle, stylohyoid muscle, stapedius muscle	Lesser horn and upper body of hyoid bone, stapes, styloid process, stylohyoid ligament
3	CN IX	Stylopharyngeus muscle	Greater horn and lower body of hyoid bone
4	CN X (superior laryngeal branch)	Muscles of the soft palate (except tensor veli palatini), muscles of the pharynx (except stylopharyngeus), cricothyroid muscle, cricopharyngeus muscle, laryngeal cartilages	. . .
6	CN X (recurrent laryngeal branch)	Intrinsic muscles of the larynx (except cricothyroid), upper muscles of esophagus, laryngeal cartilages	. . .

Table 9-2
Adult Derivatives of the Pharyngeal Pouches

Pouch	Adult Derivatives
1	Epithelial lining of auditory tube and middle ear cavity
2	Epithelial lining of palatine tonsil crypts
3	Inferior parathyroid gland, thymus
4	Superior parathyroid gland, ultimobranchial body*

* Neural crest cells migrate into the ultimobranchial body to form the parafollicular cells of the thyroid.

B. The **secondary palate** forms from outgrowths of the maxillary prominences called **palatine shelves.** These palatine shelves fuse at the midline.

C. The **definitive palate** is formed following fusion of the primary and secondary palates at the **incisive foramen.**

V. CLINICAL CORRELATIONS

A. A **pharyngeal fistula** occurs when pouch 2 and groove 2 persist. The fistula is generally found along the anterior border of the sternocleidomastoid muscle.

B. A **pharyngeal cyst** occurs when pharyngeal grooves that are normally obliterated persist. The cyst is usually located at the angle of the mandible.

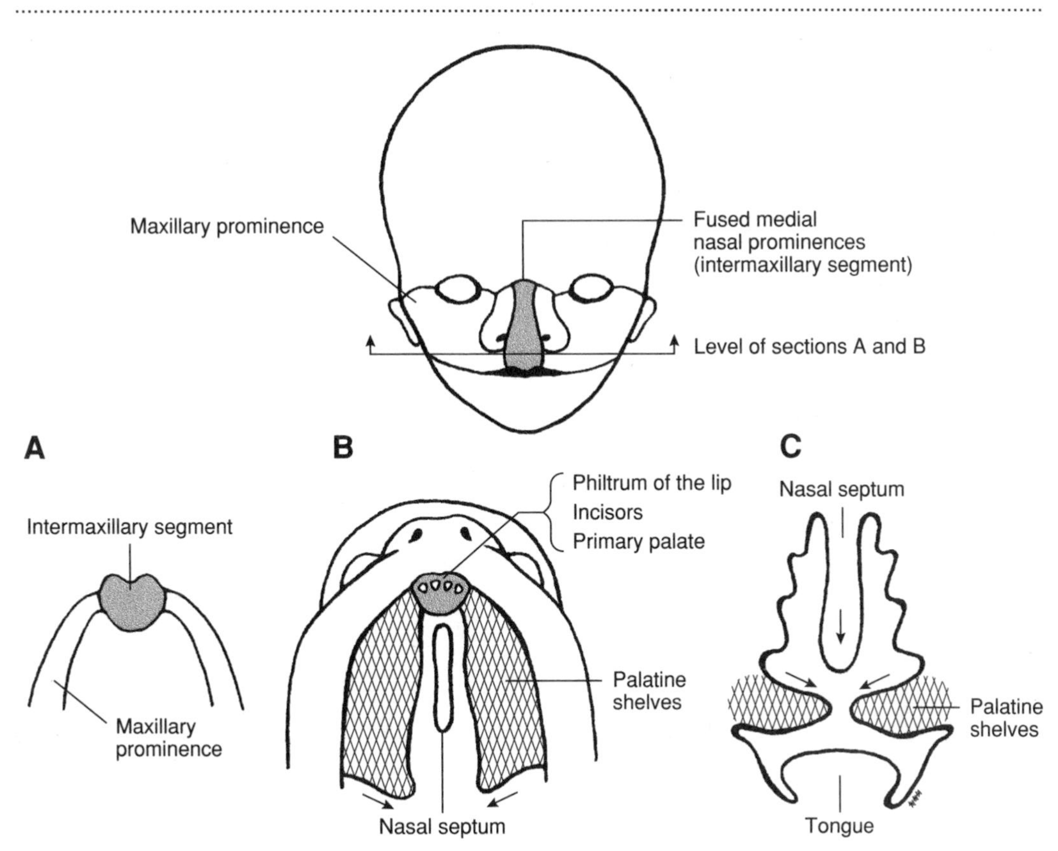

Figure 9-2. Formation of the palate. (*A*,*B*) The ventral view with the lower jaw and tongue removed shows the intermaxillary segment and palatine shelves (*crosshatched area*) fusing at the midline (*arrows*). (C) The frontal view shows the palatine shelves (*crosshatched area*) fusing at the midline with the nasal septum (*arrows*).

C. **First arch syndrome,** which is characterized by facial anomalies, occurs when faulty migration of neural crest cells causes abnormal formation of pharyngeal arch 1. Two well described syndromes are **Treacher Collins syndrome** and **Pierre Robin syndrome.**

D. **DiGeorge syndrome** occurs when pharyngeal pouches 3 and 4 fail to differentiate into the parathyroid glands and thymus.

E. **Ectopic thyroid, parathyroid,** or **thymus** results from abnormal migration of these glands from their embryonic position to their adult anatomical position.

F. A **thyroglossal duct cyst** occurs when parts of the thyroglossal duct persist, generally at the midline near the hyoid bone. The cyst may also be found at the base of the tongue **(lingual cyst)**.

G. **Cleft palate** occurs when the palatine shelves fail to fuse with each other or the primary palate.

H. **Cleft lip** occurs when the maxillary prominence fails to fuse with the medial nasal prominence. Cleft palate and cleft lip are distinct malformations, although they often occur together.

I. **Ankyloglossia** occurs when the tongue is not freed from the floor of the mouth. Most commonly, the frenulum extends to the tip of the tongue.

J. **Encephalocele (cranial meningocele)** occurs when brain tissue, meninges, or both herniate through defects in the skull.

K. **Craniosynostosis** occurs as a result of premature closure of one or more sutures of the skull.

10

Urinary System

I. OVERVIEW. **Intermediate mesoderm** forms a longitudinal elevation along the dorsal body wall, called the **urogenital ridge** (Figure 10-1). A portion of the urogenital ridge, the **nephrogenic cord,** forms the pronephros, mesonephros, and metanephros.

A. The **pronephros** completely regresses.

B. The **mesonephros** forms the **mesonephric (wolffian) duct.**

C. The **metanephros** develops from the **ureteric bud,** an outgrowth of the mesonephric duct, and **metanephric mesoderm** (Table 10-1). The metanephros **eventually becomes the definitive adult kidney.**

1. The permanent kidney **ascends** during development from the sacral region to its adult anatomical location at T12–L3 (Figure 10-2).
2. Embryonic arteries formed during the ascent may persist as **supernumerary arteries** in the adult. These arteries are **end arteries;** therefore damage to the vessels will result in damage to the tissues they supply.

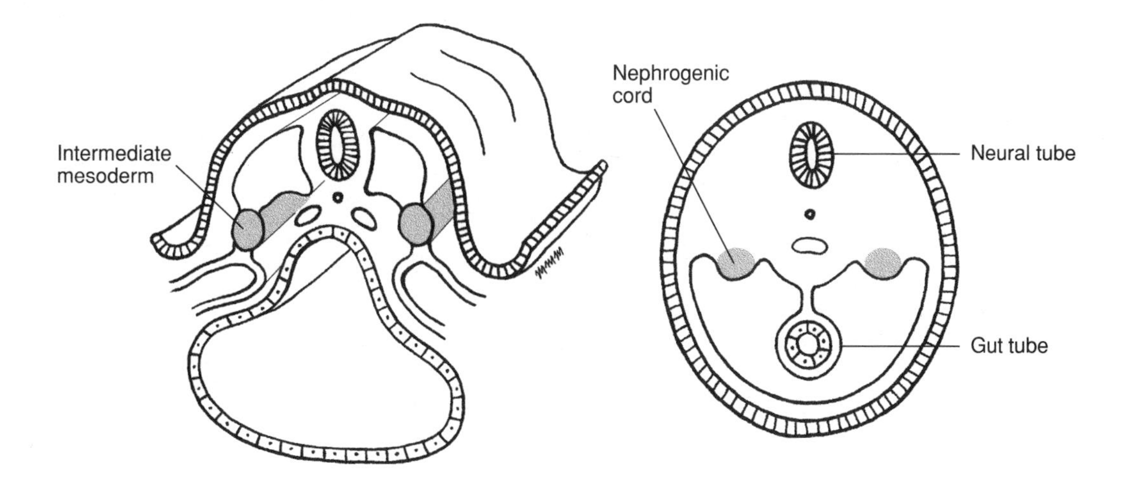

Figure 10-1. Formation of the nephrogenic cord as the embryo goes through craniocaudal and lateral folding.

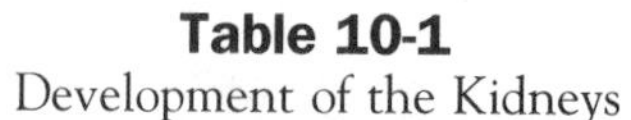

Table 10-1
Development of the Kidneys

Embryonic Structure	Adult Derivative	
Ureteric bud	Collecting duct	
	Minor calyces	
	Major calyces	
	Renal pelvis	
	Ureter	
Metanephric mesoderm	Renal glomerulus (capillaries)	
	Renal (Bowman's) capsule	
	Proximal convoluted tubule	
	Proximal straight tubule	Loop of Henle
	Descending thin limb	Loop of Henle
	Ascending thin limb	Loop of Henle
	Distal straight tubule	Loop of Henle
	Distal convoluted tubule	
	Connecting tubule (CT)	

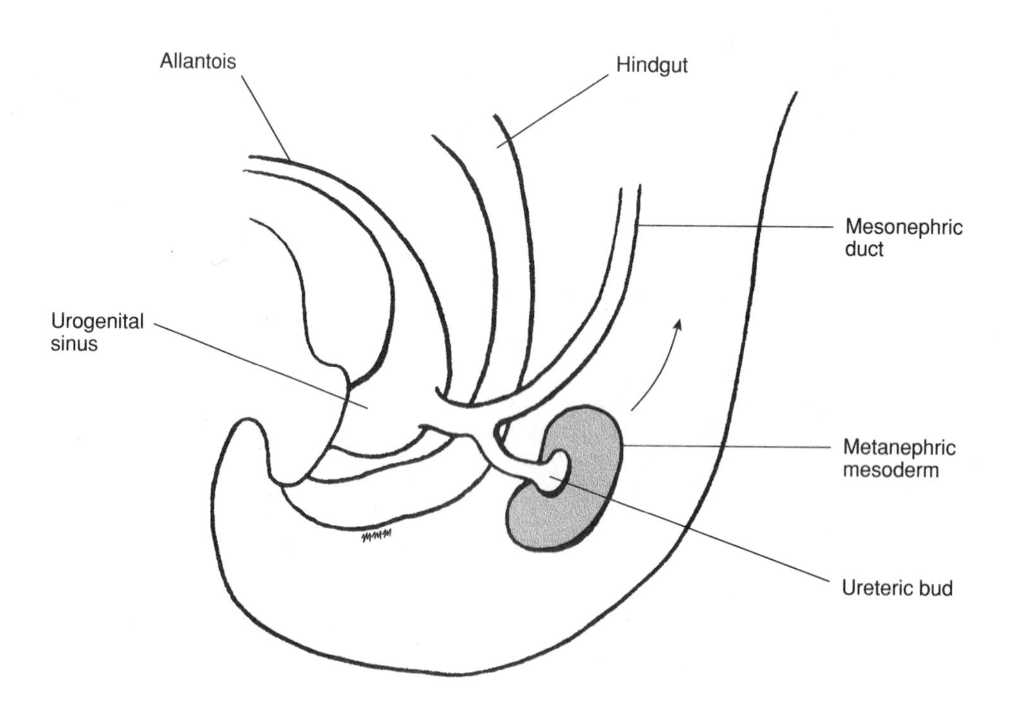

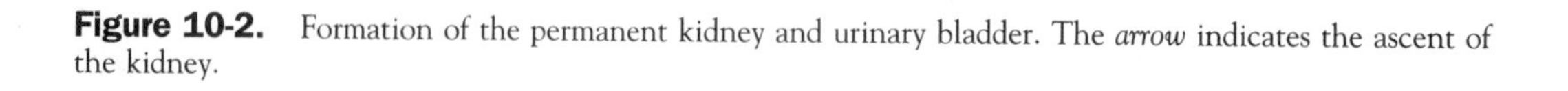

Figure 10-2. Formation of the permanent kidney and urinary bladder. The *arrow* indicates the ascent of the kidney.

II. SUPRARENAL GLAND

A. The **cortex** forms as a result of two episodes of **mesoderm** proliferation.

1. The first episode forms the **fetal cortex,** which regresses by the second postnatal month.
2. The second episode forms the **adult cortex** (i.e., the zona glomerulosa, zona fasciculata, and zona reticularis).

B. The **medulla** forms from **neural crest cells,** which migrate to the fetal cortex and differentiate into **chromaffin cells.**

III. URINARY BLADDER.

The urinary bladder develops from the upper end of the **urogenital sinus,** which is continuous with the allantois.

A. The allantois degenerates and forms a fibrous cord in the adult called the **urachus (median umbilical ligament).**

B. The **trigone of the bladder** is formed by the incorporation of the lower end of the mesonephric ducts into the posterior wall of the urogenital sinus.

IV. CLINICAL CORRELATIONS

A. Renal agenesis occurs when the ureteric bud fails to develop.

B. Horseshoe kidney occurs when the inferior poles of both kidneys fuse. During the ascent, the horseshoe kidney gets trapped behind the inferior mesenteric artery.

C. Nephroblastoma (Wilm's tumor) is a common malignant tumor found in children. The neoplasm probably arises from embryonic nephrogenic tissue.

D. Urachal cyst (sinus) occurs when a remnant of the allantois persists. This condition is associated with urine drainage from the umbilicus.

E. Pheochromocytoma is a chromaffin cell tumor. Pheochromocytomas are generally found along the migratory path of neural crest cells (e.g., to the fetal cortex).

11

Reproductive System

I. INDIFFERENT EMBRYO

I. INDIFFERENT EMBRYO (Figure 11-1). Although the genotype of the embryo is established at fertilization, female and male embryos are phenotypically indistinguishable between weeks 1 and 6.

A. By week 12, some female and male characteristics of the external genitalia can be recognized.

B. By week 20, phenotypic differentiation is complete. The components of the indifferent embryo that are remodeled to form the adult female and male reproductive systems are the **gonads, paramesonephric (müllerian) ducts, mesonephric (wolffian) ducts** and **tubules, urogenital sinus, phallus, urogenital folds,** and **labioscrotal swellings** (Table 11-1).

II. DESCENT OF THE OVARIES AND TESTES. The ovaries and testes develop within the abdominal cavity but later descend into the pelvis and scrotum, respectively. The **gubernaculum** (a band of fibrous tissue) and the **processus vaginalis** (an evagination of peritoneum) are involved in the descent of both the ovaries and testes. The fates of the gubernaculum and processus vaginalis are summarized in Table 11-2.

III. CLINICAL CORRELATIONS

A. Female pseudointersexuality occurs when ovarian tissue is present in a person with a normal female karyotype (46,XX), but the external genitalia have undergone masculinization. This condition is commonly caused by **congenital adrenal hyperplasia,** which results in excess production of androgens by the fetus.

B. Male pseudointersexuality occurs when testicular tissue is present in a person with a normal male karyotype (46,XY), but development of the male external genitalia is stunted. This condition is commonly caused by inadequate production of testosterone and müllerian-inhibiting factor (MIF).

C. Testicular feminization syndrome occurs when a 46,XY fetus develops **testes** and **female** (instead of male) **external genitalia.** This condition is caused by a **lack of androgen receptors** in the urogenital folds and labioscrotal swellings. These individuals are considered female medically, legally, and socially.

D. Hypospadias occurs when the urogenital folds fail to fuse completely. As a result, the urethra opens onto the ventral surface of the penis.

Table 11-1
Development of the Adult Female and Male Reproductive Systems

Indifferent Embryo	Adult Female	Adult Male
Gonads	Ovary, follicles, rete ovarii	Testes, seminiferous tubules, rete testes
Paramesonephric ducts	Uterine tubes, uterus, cervix, and upper part of vagina	Appendix of testes (vestigial)
Mesonephric ducts	Duct of Gartner (vestigial)	Epididymis, ductus deferens, seminal vesicles, ejaculatory duct
Mesonephric tubules	Epoophoron, paroophoron (vestigial)	Efferent ductules
Urogenital sinus	Urinary bladder, urethra, urethral and paraurethral glands, greater vestibular glands, lower part of vagina	Urinary bladder, urethra, prostate gland, bulbourethral glands
Phallus	Clitoris	Penis
Urogenital folds	Labia minora	Ventral aspect of penis Penile raphe
Labioscrotal swellings	Labia majora	Scrotum Scrotal raphe

Table 11-2
Fate of the Gubernaculum and Processus Vaginalis

Indifferent Embryo	Adult Female	Adult Male
Gubernaculum	Ovarian ligament, round ligament of the uterus	Gubernaculum testes
Processus vaginalis	. . .	Tunica vaginalis

E. **Cryptorchidism** occurs when the testes fail to descend into the scrotum. Bilateral cryptorchidism may result in sterility.

F. **Hydrocele of the testes** occurs when a small patency of the tunica vaginalis remains so that peritoneal fluid can flow into it, resulting in a fluid-filled cyst near the testes.

G. **Congenital inguinal hernia** occurs when a large patency of the tunica vaginalis remains so that a loop of intestine herniates into the scrotum.

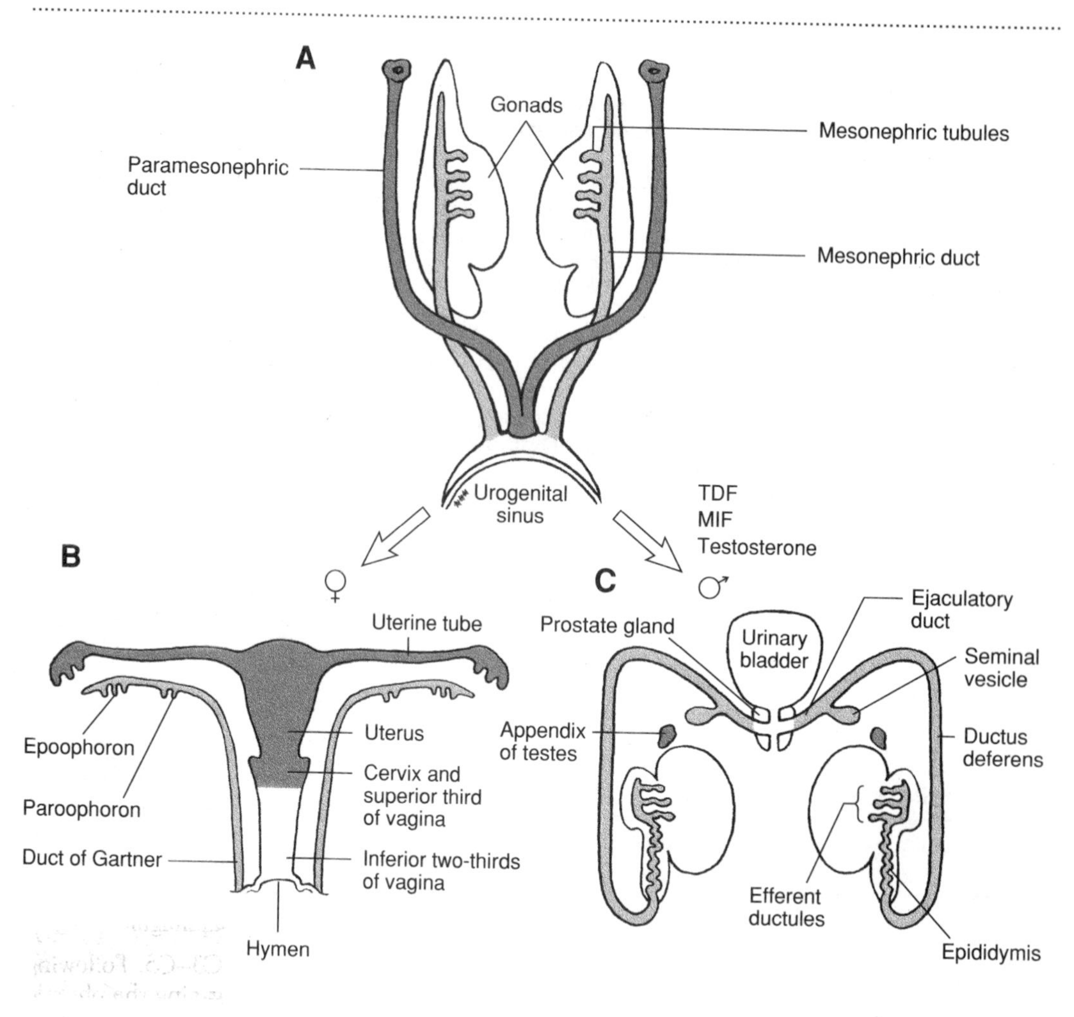

Figure 11-1. (*A*) The indifferent embryo. The paramesonephric ducts play a major role in female development. The mesonephric ducts and tubules play a major role in male development. (*B*) Adult female structures form from the paramesonephric duct and vestigial remnants of the mesonephric ducts and tubules. (*C*) Adult male structures form from the mesonephric ducts and tubules and vestigial remnants of the paramesonephric ducts. The production of **testes-determining factor** (*TDF*), **müllerian-inhibiting factor** (*MIF*), and **testosterone** direct the indifferent embryo toward male development. It is believed that TDF (a 220-amino acid nonhistone protein) is the gene product of the sry gene, which is located on the short arm of the Y chromosome.

12

Body Cavities

I. INTRAEMBRYONIC COELOM.

The intraembryonic coelom is initially one continuous space. The formation of the **pleuropericardial membranes** and the **diaphragm** partitions this space into the pericardial, pleural, and peritoneal cavities of the adult.

A. **Pleuropericardial membranes** are sheets of mesoderm that separate the pleural cavity from the pericardial cavity and later form the **fibrous pericardium.** This relationship is evidenced by the fact that, to reach the diaphragm, the **phrenic nerves** course through the pleuropericardial membranes in the embryo and the fibrous pericardium in the adult.

B. **Diaphragm.** The diaphragm separates the pleural cavity from the peritoneal cavity.

1. **Formation.** It is formed by the fusion of tissue from four sources.
   - **a.** The **septum transversum** gives rise to the **central tendon of the diaphragm** in the adult.
   - **b.** The **pleuroperitoneal membranes** contribute tissue to the diaphragm.
   - **c.** The **dorsal mesentery of the esophagus** gives rise to the **crura of the diaphragm** in the adult.
   - **d.** The **body wall** contributes muscle to the periphery of the diaphragm.
2. **Descent.** The septum transversum initially lies at the level of C3–C5. Following the rapid growth of the neural tube, the diaphragm descends, carrying the phrenic nerves along with it.

II. CLINICAL CORRELATIONS

A. **Congenital diaphragmatic hernia** is a protrusion of the abdominal contents into the pleural cavity following failure of the **pleuroperitoneal membranes** to develop properly. The hernia is most commonly found on the **left posterolateral side** and causes **pulmonary hypoplasia.**

B. **Esophageal hiatal hernia** is the protrusion of the stomach into the pleural cavity through an abnormally large esophageal hiatus. This condition renders the **esophagogastric sphincter** incompetent, causing the reflux of stomach contents into the esophagus.

13

Nervous System

I. NEURAL TUBE

A. Formation (Figure 13-1)

1. The **notochord** induces the overlying ectoderm to differentiate into **neuroectoderm** to form the **neural plate.** In the adult, the notochord forms the **nucleus pulposus** of the intervertebral disk.
2. The neural plate folds to give rise to the **neural tube.**
   a. As the neural plate folds, some cells differentiate into pluripotent **neural crest cells.**
   b. The neural tube is initially connected to the amniotic cavity via the **anterior** and **posterior neuropores.** The **lamina terminalis** marks the location of the anterior neuropore in the adult.

B. Vesicles. The neural tube develops **three primary vesicles** and **five secondary vesicles.** These vesicles give rise to various adult structures (Table 13-1).

C. Cells of the neural tube wall give rise to the following cells of the central nervous system (CNS).

1. **Neuroblasts** form all neurons within the brain and spinal cord, including the preganglionic sympathetic and preganglionic parasympathetic neurons.
2. **Glioblasts** are the supporting cells within the CNS.
   a. **Astrocytes** surround capillaries.
   b. **Oligodendrocytes** produce myelin.
   c. **Ependymocytes** line the ventricles and central canal.
   d. **Tanycytes** line the third ventricle and transport substances from the cerebrospinal fluid (CSF) to the hypothalamic portal system.
   e. **Choroid plexus cells** produce CSF. The tight junctions between them form the blood–CSF barrier.
   f. **Microglia,** phagocytic cells of the CNS, are derived from monocytes.

II. POSITIONAL CHANGES OF THE SPINAL CORD

A. At week 8 of development, the spinal cord extends the entire length of the bony vertebral canal.

B. At birth, the conus medullaris extends to the L3 vertebra.

C. In the adult, the conus medullaris extends to the L1 vertebra.

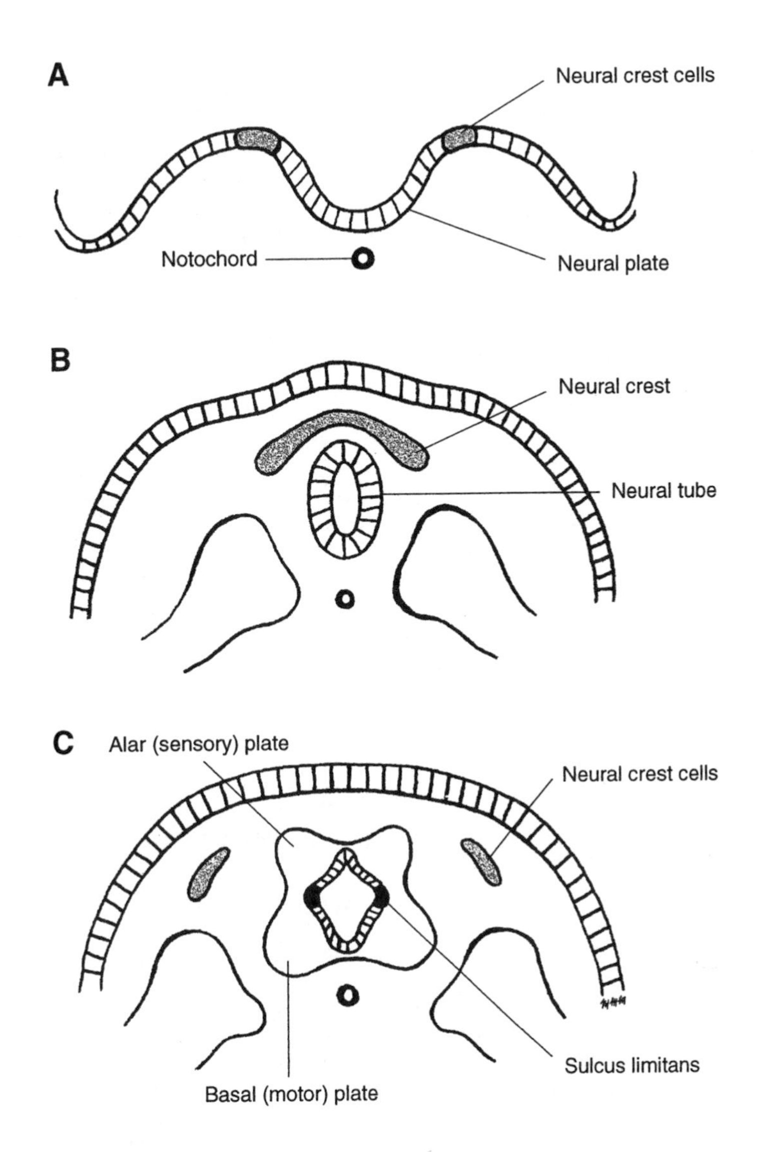

Figure 13-1. Stages in the development of the neural tube and neural crest cells. (*A*) The notochord induces the overlying ectoderm to form the neural plate. (*B*) The neural plate folds to give rise to the neural tube. As the neural plate folds, it gives rise to neural crest cells. (*C*) The **sulcus limitans** is a groove in the lateral wall of the ventricle that separates the alar (sensory) plate from the basal (motor) plate. The alar plate becomes the **dorsal horn** of the spinal cord and the basal plate becomes the **ventral horn** of the spinal cord.

Table 13-1
Development of the Brain from the Neural Tube*

Primary Vesicles	Secondary Vesicles	Adult Derivatives
Prosencephalon	Telencephalon	Cerebral hemispheres, basal ganglia, lamina terminalis, olfactory bulbs, hippocampus
	Diencephalon	Epithalamus, thalamus, hypothalamus, neurohypophysis, pineal gland, retina, optic nerve, mamillary bodies
Mesencephalon	Mesencephalon	Midbrain
Rhombencephalon	Metencephalon	Pons, cerebellum
	Myelencephalon	Medulla

***The remainder of the neural tube forms the spinal cord.**

Table 13-2
Origination of the Sympathetic Nervous System

Embryonic Structure	Adult Derivative
Basal plate of neural tube	Preganglionic sympathetic neurons within the intermediolateral cell column
Neural crest cells	Postganglionic sympathetic neurons within the sympathetic chain ganglia and prevertebral ganglia

Table 13-3
Origination of the Parasympathetic Nervous System

Embryonic Structure	Adult Derivative
Basal plate of neural tube	Preganglionic parasympathetic neurons within the nuclei of the midbrain (III), pons (VII), and medulla (IX, X)
	Preganglionic parasympathetic neurons within the spinal cord nucleus at S2–S4
Neural crest cells	Postganglionic parasympathetic neurons within the ciliary (III), pterygopalatine (VII), submandibular (VII), otic (IX), and enteric (X) ganglia
	Postganglionic parasympathetic neurons within the ganglia of the abdominal and pelvic cavities

III. MENINGES

A. The **dura mater** arises from mesoderm that surrounds the neural tube.

B. The **pia mater** and **arachnoid membrane** arise from neural crest cells.

IV. AUTONOMIC NERVOUS SYSTEM

A. The **sympathetic nervous system** originates from the basal plate of the neural tube and neural crest cells (Table 13-2).

B. The **parasympathetic nervous system** also originates from the basal plate of the neural tube and neural crest cells (Table 13-3).

V. HYPOPHYSIS

A. The **adenohypophysis** develops from an evagination of ectoderm from the roof of the primitive mouth **(Rathke's pouch).**

B. The **neurohypophysis** develops from an evagination of neuroectoderm from the diencephalon.

VI. CLINICAL CORRELATIONS

A. Spina bifida

1. **Spina bifida occulta** occurs when there is only a defect of the vertebral arches.
2. **Spina bifida with meningocele** occurs when the meninges project through a vertebral defect.
3. **Spina bifida with meningomyelocele** occurs when the meninges and spinal cord project through a vertebral defect.
4. **Spina bifida with myeloschisis** occurs when the neural tube fails to close, resulting in an open neural tube on the surface of the back. Newborn infants are paralyzed distal to the lesion.

B. **Anencephaly** occurs when the anterior neuropore fails to close, resulting in failure of the brain to develop. Generally, only a rudimentary brain stem is present.

C. **Arnold-Chiari malformation** occurs when parts of the cerebellum herniate through the foramen magnum.

D. **Dandy-Walker syndrome** is a congenital hydrocephalus associated with atresia of the foramen of Luschka and foramen of Magendie.

E. **Hydrocephalus** is most commonly caused by stenosis of the cerebral aqueduct during development. Excessive CSF accumulates in the ventricles and subarachnoid space.

F. **Fetal alcohol syndrome** is the most common cause of mental retardation. It includes microcephaly and congenital heart disease.

G. **Craniopharyngioma** is a congenital cystic tumor resulting from remnants of Rathke's pouch.

14

Ear

I. EMBRYOLOGIC ORIGINS. Table 14-1 summarizes the embryologic origins of the internal ear, middle ear, and external ear.

II. FORMATION. Figure 14-1 depicts the formation of the structures of the adult internal ear.

Table 14-1
Derivation of the Structures of the Ear

Embryonic Structure	Adult Derivatives
Otic vesicle*	Internal ear
Utricular portion	Utricle, semicircular ducts, vestibular ganglion of CN VIII
Saccular portion	Saccule, cochlear duct (organ of Corti), spiral ganglion of CN VIII
	Middle ear
Pharyngeal arch 1	Incus, malleus, tensor tympani muscle
Pharyngeal arch 2	Stapes, stapedius muscle
Pharyngeal pouch 1	Epithelial lining of auditory tube and middle ear cavity
Pharyngeal membrane 1	Tympanic membrane
	External ear
Pharyngeal groove 1	Epithelial lining of external auditory meatus
Auricular hillocks	Auricle

*Derived from surface ectoderm.

Figure 14-1. Schematic transverse sections showing the formation of the otic placode and otic vesicle from surface ectoderm. (A) The otic placode invaginates into the mesoderm and becomes the otic vesicle. (B) The vestibular and spiral ganglia are derived from the otic vesicle. (C) The adult ear. Pharyngeal arch 1 and pharyngeal arch 2 form the stapes (*St*), incus (*I*), and malleus (M). *Ut* = utricle; *Sac* = saccule.

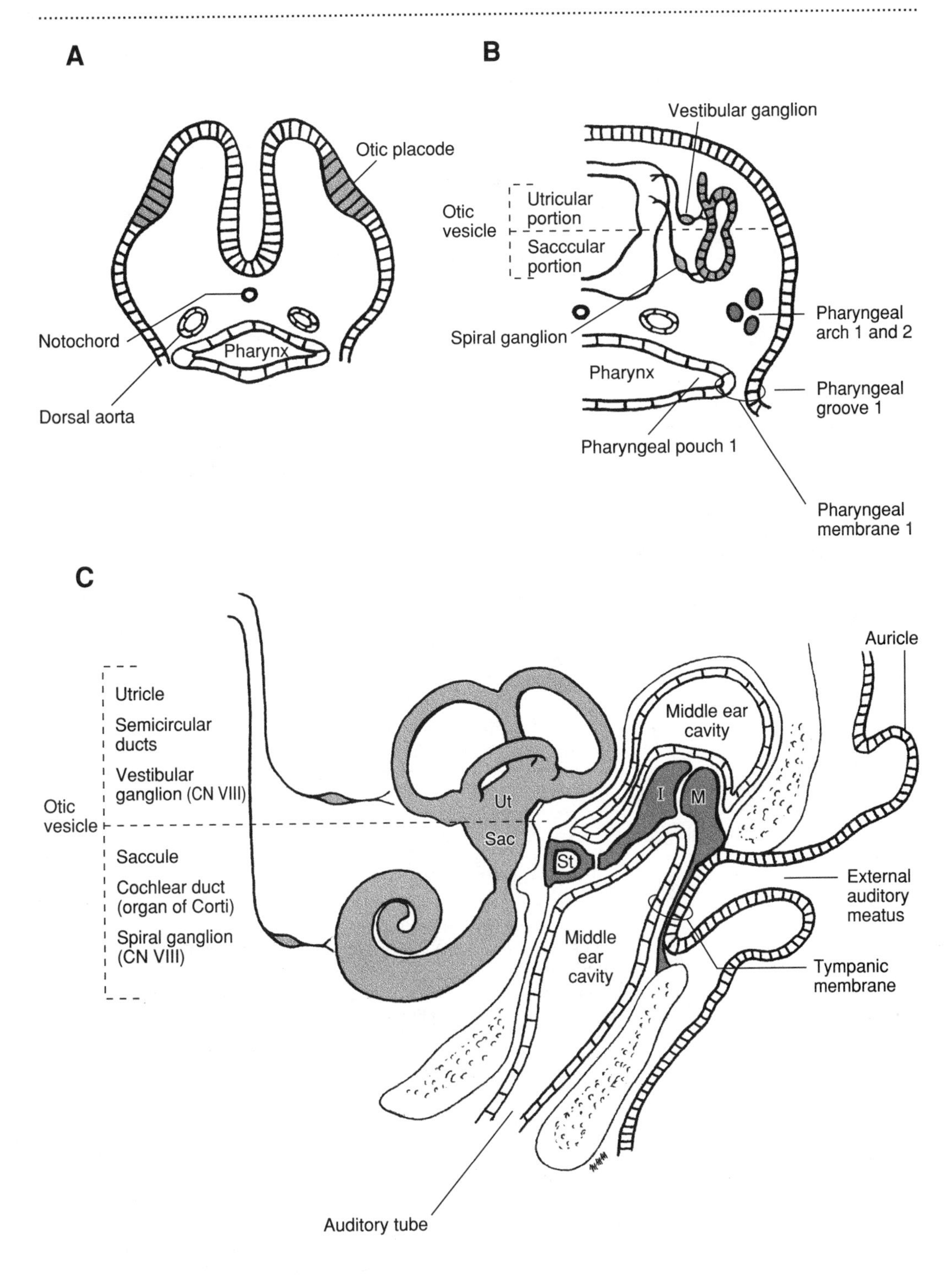
A
Otic placode
Notochord
Pharynx
Dorsal aorta
B
Vestibular ganglion
Otic vesicle
Utricular portion
Sacccular portion
Spiral ganglion
Pharyngeal arch 1 and 2
Pharynx
Pharyngeal groove 1
Pharyngeal pouch 1
Pharyngeal membrane 1
C
Utricle
Semicircular ducts
Vestibular ganglion (CN VIII)
Otic vesicle
Saccule
Cochlear duct (organ of Corti)
Spiral ganglion (CN VIII)
Ut
Sac
St
I
M
Middle ear cavity
Middle ear cavity
Auricle
External auditory meatus
Tympanic membrane
Auditory tube

15

Eye

I. EYE. The eye is formed in part from a neuroectodermal evagination of the **diencephalon** called the **optic cup** and **optic stalk** (Figure 15-1). In addition, surface ectoderm (the **lens placode**), mesoderm, and neural crest cells contribute to the formation of a number of structures of the adult eye (Table 15-1).

II. CLINICAL CORRELATIONS

A. Coloboma iridis occurs when the choroid fissure fails to close, causing a cleft in the iris.

B. Persistent iridopupillary membrane occurs when strands of connective tissue cover the pupil.

Table 15-1
Derivation of the Structures of the Eye

Embryonic Structure	Adult Derivative
Neuroectoderm (diencephalon)	
Optic cup	Retina, iris, ciliary body
Optic stalk	Optic nerve (CN II)
Surface ectoderm	
Lens placode	Lens, anterior epithelium of cornea
Mesoderm	Sclera, substantia propria of cornea, corneal endothelium, vitreous body, extraocular muscles
Hyaloid artery and vein	Central artery and vein of retina (branch of the ophthalmic artery)
Neural crest cells	Choroid, sphincter pupillae muscle, dilator pupillae muscle, ciliary muscle

Figure 15-1. (*A*) The optic cup and optic stalk are evaginations of the diencephalon. The optic cup induces surface ectoderm to differentiate into the lens placode. (*B*) Formation of the optic nerve (CN II) from the optic stalk. The choroid fissure, which is located on the undersurface of the optic stalk, permits access of the hyaloid artery and vein to the inner aspect of the eye. The choroid fissure eventually closes. As ganglion cells form in the retina, axons accumulate in the optic stalk and cause the inner and outer layers of the optic stalk to fuse, obliterating the lumen and forming the optic nerve. (*C*) The adult eye. Note that the sclera is continuous with the dura and the choroid is continuous with the pia–arachnoid. The iridopupillary membrane is normally obliterated.

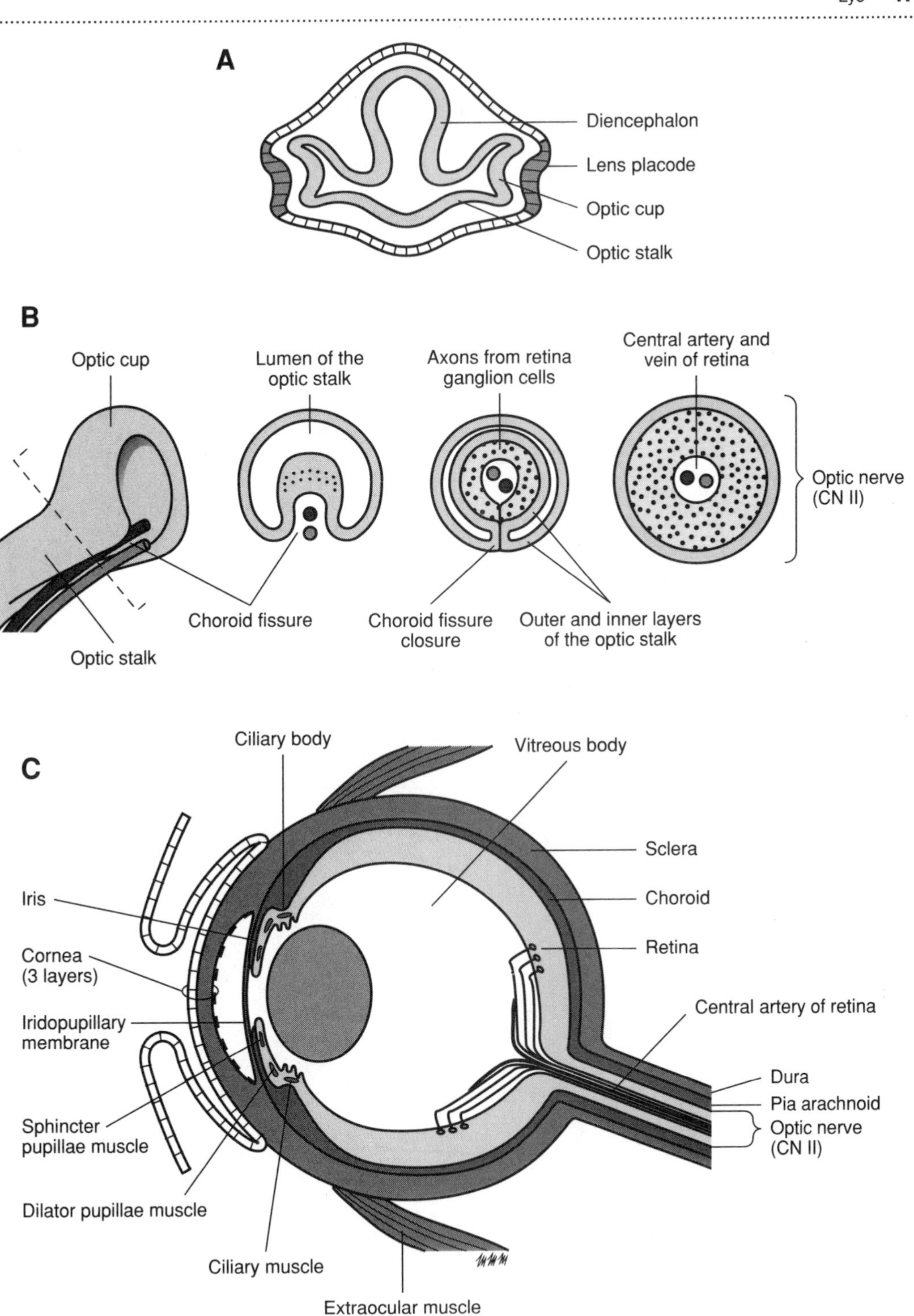
A
Diencephalon
Lens placode
Optic cup
Optic stalk
B
Optic cup
Lumen of the optic stalk
Axons from retina ganglion cells
Central artery and vein of retina
Optic nerve (CN II)
Choroid fissure
Choroid fissure closure
Outer and inner layers of the optic stalk
Optic stalk
C
Ciliary body
Vitreous body
Sclera
Choroid
Retina
Iris
Cornea (3 layers)
Iridopupillary membrane
Central artery of retina
Dura
Pia arachnoid
Optic nerve (CN II)
Sphincter pupillae muscle
Dilator pupillae muscle
Ciliary muscle
Extraocular muscle